# SpringerBriefs in Food, Health, and Nutrition

SpringerBriefs in Food, Health, and Nutrition present concise summaries of cutting edge research and practical applications across a wide range of topics related to the field of food science.

Editor-in-Chief

Richard W. Hartel, University of Wisconsin—Madison, USA

Associate Editor

J. Peter Clark, Consultant to the Process, Industries, USA
John W. Finley, Louisiana State University, USA
David Rodriguez-Lazaro, ITACyL, Spain
David Topping, CSIRO, Australia

For further volumes:
http://www.springer.com/series/10203

C. Anandharamakrishnan

# Computational Fluid Dynamics Applications in Food Processing

 Springer

C. Anandharamakrishnan
CSIR-Central Food Technological
    Research Institute
Mysore
India

ISBN 978-1-4614-7989-5          ISBN 978-1-4614-7990-1    (eBook)
DOI 10.1007/978-1-4614-7990-1
Springer New York Heidelberg Dordrecht London

Library of Congress Control Number: 2013940764

Printed on acid-free paper

Springer is part of Springer Science+Business Media (www.springer.com)

*Dedicated to my Parents*

# Acknowledgments

I am extremely grateful to Prof. Ram Rajasekharan, Director, CSIR-Central Food Technological Research Institute, Mysore, India for his valuable guidance, scientific advice and continuous encouragement.

I would like to express my sincere gratitude to my guide and mentor Prof. Chris Rielly, Professor and Head, Chemical Engineering Department, Loughborough University, UK for his never ending inspiration, guidance and support. He has helped me to understand the concepts of CFD modelling. I sincerely thank Dr. Andy Stapley, Senior Lecturer, Chemical Engineering Department, Loughborough University, UK for his help and support.

I gratefully acknowledge the Commonwealth Scholarship Commission, UK and Department of Science and Technology, Government of India for the financial support, which enabled some of the works presented in this book to be carried out.

I would like to thank all my Ph.D. students and especially Mr. Chhanwal, Mr. Gopirajah and Ms. Padma Ishwarya for their help.

My heartfelt thanks to my parents and sister for their prayers, love, encouragement and support right from the beginning.

This work would not have been possible without my wife Dr. G. Shashikala and my son A. Nishanth, I appreciate their sacrifice, patience and moral support throughout my research career.

# Contents

# Chapter 1
# Computational Fluid Dynamics Applications in Food Processing

Computational Fluid Dynamics (CFD) has been extensively applied in the food-processing sector for the design and optimization of equipment such as ovens, spray dryers, chillers, heat exchangers, etc. Numerous benefits from the implementation of CFD models have been reported. This book recapitulates the application of CFD modeling; in particular, design and optimization of spray drying, spray freezing, baking ovens, high pressure processing, retorts processing and also biological systems. CFD modeling is often used in spray drying operations, as it is very difficult and expensive to obtain measurements of airflow, temperature, particle size, and humidity within the drying chamber. CFD can be a useful tool for predicting the gas flow pattern and particle histories such as temperature, velocity, residence time, and impact position during spray drying. CFD modeling of a baking oven provides constructive information about temperature and airflow pattern throughout the baking chamber to enhance heat transfer, and in turn, final product quality. CFD modeling also helps in designing the ovens for rapid bread baking. CFD modeling can be used in retort processing of canned solid and liquid foods for understanding and optimization of the heat transfer processes. CFD can be used to numerically model the dynamics of gastrointestinal contents during digestion, based on the motor response of the gastrointestinal (GI) tract and the physicochemical properties of luminal contents. Advanced computational fluid dynamics programs offer a promising technique to characterize the mechanisms promoting digestion. Furthermore, this book predominantly focuses on the recent developments in this field, constraints in CFD modeling approaches, their strengths and limitations, and future applications in food industries.

## 1.1 Introduction to Computational Fluid Dynamics

CFD is a simulation tool that uses powerful computers in combination with applied mathematics to model fluid flow situations and aid in the optimal design of industrial processes. The method comprises solving equations for the conservation of mass, momentum and energy, using numerical methods to give predictions of

C. Anandharamakrishnan, *Computational Fluid Dynamics Applications*
*in Food Processing*, SpringerBriefs in Food, Health, and Nutrition,
DOI: 10.1007/978-1-4614-7990-1_1, © Chinnaswamy Anandharamakrishnan 2013

velocity, temperature and pressure profiles inside the system. Its powerful graphics can be used to show the flow behaviour of fluid with three dimensional (3D) images (Anderson 1984; Scott and Richardson 1997).

The history of CFD takes us way back to the 1960s, when the aerospace industry integrated this technique into the design, research and development, and manufacture of aircraft and jet engines. Around the 1970s, CFD became an acronym for a combination of physics, numerical mathematics, and, to some extent, computer sciences employed to simulate fluid flows. However, the applications were mostly restricted to the two-dimensional (2D) flow models due to the low speed and storage capacities of the computers. The beginning of CFD was triggered by the availability of more powerful mainframes, and the advances in CFD are still tightly coupled to the evolution of computer technology. Around the mid-1980s, computer predictions of fluid flow have been used routinely in both science and engineering to produce results. With the advances of numerical methodologies, particularly of implicit schemes, the solution of flow problems that require real gas modeling became feasible by the end of 1980s. Toward the 1990s, 3D modeling became possible and led to an upsurge of interest in a great deal of industrial applications. Nowadays, CFD methodologies are routinely employed in the fields of aircraft, turbo machinery, car, and ship design. Furthermore, CFD is also applied in meteorology, oceanography, astrophysics, and also in architecture (Anderson 1984; Shaw 1992; Versteeg and Malalasekera 1995; Blazek 2001). For more a detailed historical perspective, the books by Roache (1976) and Tannehill et al. (1997) are highly recommended. Today, CFD finds extensive usage in basic and applied research, in design of engineering equipment and in calculation of environmental and geophysical phenomena.

## 1.2  Theory of CFD Modeling

CFD is a numerical technique for the solution of equations governing the flow of fluids inside defined flow geometry. The flow of any fluid can be described using the following transport Eqs. (1.1–1.4) (Bird et al. 1960; Versteeg and Malalasekera 1995; Marshall and Bakker 2002; Fluent 2006). These equations are derived by considering mass, momentum and energy balances in an element of fluid, resulting in a set of partial differential equations. They are completed by adding other algebraic equations from thermodynamics, such as the equation of state for density and a constitutive equation to describe the rheology (Fletcher 2000).

### 1.2.1  Conservation of Mass Equation

The continuity equation describes the rate of change of density at a fixed point resulting from the divergence in the mass velocity vector $\rho v$. Equation (1.1) is the unsteady, three-dimensional, mass conservation or continuity equation for the simplified case of a constant density fluid (incompressible fluid).

$$\nabla \cdot v = 0 \tag{1.1}$$

where $\nabla$ has the dimension of reciprocal length:

$$\nabla = \frac{\partial}{\partial x}i + \frac{\partial}{\partial y}j + \frac{\partial}{\partial z}k \tag{1.2}$$

### 1.2.2 Momentum Equation

The principles of the conservation of momentum is an application of Newton's second law of motion to an element of fluid, and states that a small volume of element moving with the fluid is accelerated because of the force acting upon it.

$$\rho_g \frac{Dv}{Dt} = -\nabla p + \nabla \cdot \underline{\underline{\tau}} + \rho_g \underline{g} \tag{1.3}$$

In Eq. (1.3), the convection terms are on the left side, and on the right hand side are the pressure gradient ($p$), source terms of gravitational force ($\underline{g}$) and stress tensor ($\underline{\underline{\tau}}$), which is responsible for diffusion of momentum.

### 1.2.3 Energy Equation

The first law of thermodynamics states that the rate of change of internal energy plus kinetic energy is equal to the rate of heat transfer minus the rate of work done by system. Fluent solves the energy equation in the following form.

$$\frac{\partial}{\partial t}(\rho E) + \nabla \cdot \left[ \underline{v}(\rho E + p) \right] = \nabla \cdot \left[ k_{\text{eff}}\nabla T - \sum_j h_j \underline{J}_j + (\underline{\underline{\tau}} \cdot \underline{v}) \right] \tag{1.4}$$

where $E$ is the internal (thermal) energy, $k_{\text{eff}}$ is the effective conductivity ($k_{\text{ta}} + k_{\text{t}}$, where $k_{\text{ta}}$ is thermal conductivity and $k_{\text{t}}$ is turbulent thermal conductivity), $T$ is the temperature, $\underline{\underline{\tau}}$ is stress tensor, $\underline{J}_j$ is the diffusion flux of species $j$, and $h_j$ is the enthalpy of species $j$. The three terms on the right-hand side of the equation represent energy transfer due to conduction, species diffusion and viscous dissipation, respectively.

## 1.3 Turbulence Model

There are two types of flows; namely, laminar and turbulent. Above a certain Reynolds number, all flows become unstable and exhibit turbulent behaviour. For laminar flow problems (low Reynolds number), the flows can be solved by conservation equations. In the case of turbulent flows (high Reynolds number), the computational effort involved in solving those for all time and length scales is prohibitive. An engineering approach to calculate time-averaged flow fields for turbulent flows will be developed for solving turbulent flow problems (Marshall and Bakker 2002).

The turbulence models commonly used for simulations are:

(i)   Standard $k$–$\varepsilon$ ($k$—turbulence kinetic energy and $\varepsilon$—turbulence dissipation rate)
(ii)  Renormalization Group (RNG) $k$–$\varepsilon$
(iii) Realizable $k$–$\varepsilon$
(iv)  Reynolds Stress Model (RSM)

Three models (standard, RNG, and realizable $k$–$\varepsilon$) have similar forms, with transport equations for $k$ and $\varepsilon$.

Most commercial CFD codes use turbulence models that are based on the splitting up of instantaneous quantities into a time-averaged and a fluctuating part by a process known as Reynolds decomposition. For turbulent flows, the standard $k$–$\varepsilon$ model ($k$—turbulence kinetic energy and $\varepsilon$—turbulence dissipation rate) is the most commonly used, because it converges considerably better than Reynolds stress model (RSM) (Versteeg and Malalasekera 1995), and is given as follows:

$$\frac{\partial}{\partial t}(\rho k) + \nabla \cdot (\rho k \underline{v}) = \nabla \cdot \left[\left(\mu + \frac{\mu_t}{\sigma_k}\right)\nabla k\right] + G_k - \rho \varepsilon \tag{1.5}$$

$$\frac{\partial}{\partial t}(\rho \varepsilon) + \nabla \cdot (\rho \varepsilon \underline{v}) = \nabla \cdot \left[\left(\mu + \frac{\mu_t}{\sigma_\varepsilon}\right)\nabla \varepsilon\right] + C_{l\varepsilon}\frac{\varepsilon}{k}(G_k) - C_{2\varepsilon}\rho\frac{\varepsilon^2}{k} \tag{1.6}$$

$G_k$ is the generation of kinetic energy due to the mean velocity gradients. The quantities $\sigma_k$ and $\sigma_\varepsilon$ are the turbulent Prandtl numbers for $k$ and $\varepsilon$, respectively, and $C_{l\varepsilon}$, $C_{2\varepsilon}$, are constant. The turbulent (or eddy) viscosity $\mu_t$ is calculated from $k$ and $\varepsilon$ as follows:

$$\mu_t = \rho C_\mu \frac{k^2}{\varepsilon} \tag{1.7}$$

The model constants $C_{l\varepsilon}$, $C_{2\varepsilon}$, $C_\mu$, $\sigma_k$ and $\sigma_\varepsilon$ took the following values (Launder and Spalding 1972):

$$C_{l\varepsilon} = 1.44, \ C_{2\varepsilon} = 1.92, \ C_\mu = 0.09, \ \sigma_k = 1.0 \text{ and } \sigma_\varepsilon = 1.3$$

For calculating an approximate solution of fluid flow equations, the equations have to be made discrete. For this, the flow domain is divided into number of control volumes. This is called a grid, and at each grid cell, approximate solutions for the Navier–Stokes and the continuity equations are calculated. Table 1.1, summarises the performance of the turbulence models (Zhang et al. 2007).

## 1.4 Reference Frames

Three different reference frames are widely used: the volume of fluid (VOF), Eulerian–Eulerian (EE) and Eulerian–Lagrangian (EL) models. The volume of fluid (VOF) model is designed for two or more immiscible fluids (Fig. 1.1a) by solving a

**Table 1.1** Summary of the performance of the turbulence models (Zhang et al. 2007)

| Cases | Compared items | Turbulence models | | | | | | | |
|---|---|---|---|---|---|---|---|---|---|
| | | 0-eq | RNG $k$-$\varepsilon$ | SST $k$-$\omega$ | LRN-LS | V2f-dav | RSM-IP | DES | LES |
| Natural convection | Mean temperature | B | A | A | C | A | A | C | A |
| | Mean velocity | D | B | A | B | A | B | D | B |
| | Turbulence | n/a | C | C | C | A | C | C | A |
| Forced convection | Mean velocity | C | A | C | A | A | B | C | A |
| | Turbulence | n/a | B | C | B | B | B | C | B |
| Mixed convection | Mean temperature | A | A | A | A | A | B | B | A |
| | Mean velocity | A | B | B | B | A | A | B | B |
| | Turbulence | n/a | A | D | B | A | A | B | B |
| Strong buoyancy flow | Mean temperature | A | A | A | A | A | n/c | n/a | B |
| | Mean velocity | B | A | A | A | A | n/c | n/a | A |
| | Turbulence | B | C | A | B | B | n/c | n/a | B |
| Computing time (unit) | | 1 | 2–4 | | 4–8 | | 10–20 | $10^2$–$10^3$ | |

*A* good, *B* acceptable, *C* marginal, *D* poor, *n/a* not applicable, and *n/c* not converged

**(a)**        **(b)**        **(c)**

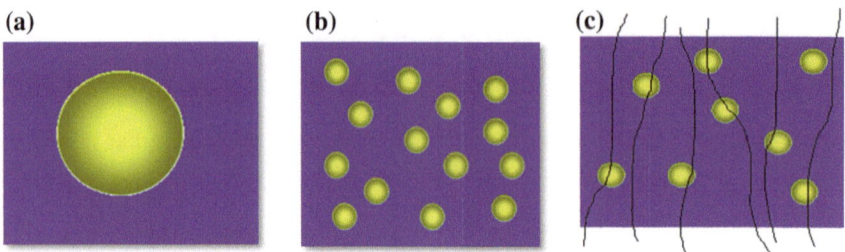

**Fig. 1.1**  Reference frames. **a** Volume of Fluid, **b** Eulerian–Eulerian, **c** Eulerian–Lagrangian

single set of momentum equations and tracking the volume fraction of each of the fluids throughout the domain. Because the fluids do not mix, each computational cell is filled with purely one fluid, purely another fluid, or the interface between two (or more) fluids. Typical applications include the prediction of jet breakup, the motion of large bubbles in a liquid, the motion of liquid after a dam break, and the steady or transient tracking of any liquid–gas interface (Fluent user guide).

The other two-phase modeling frames are the Eulerian–Eulerian and the Eulerian–Lagrangian methods. In the Eulerian–Eulerian frame (Fig. 1.1b), the dispersed phase (droplets) are treated as a continuous (Eulerian) phase, i.e. there are two Eulerian phases, one for the gas and another for droplets, which are interacting and interpenetrating continually (Mostafa and Mongia 1987). Each computational cell contains certain fractions of gas and droplets, and the transport equations are written in such a way that the volume fractions of gas and liquid sum to unity. If the computational cell consists of just a single phase, the transport equations for the two phases revert to the conventional single-phase system. The advantages of the Eulerian–Eulerian approach are usually relatively cheap in terms of computational demands for one additional set of equations, and turbulence can be modeled fairly simply. However, if a separate set of transport equations is solved for each particle size (single particle diameter was used for the dispersed phase), then the Eulerian approach can be expensive. In addition, there is some uncertainty over the most appropriate Eulerian diffusion coefficients and heat transfer coefficients. Hence, the Eulerian approach is best suited to flows with a narrow range of particle sizes where a high resolution of the particle properties is not needed (Mostafa and Mongia 1987; Jakobsen et al. 1997).

In the Eulerian–Lagrangian particle tracking approach (Fig. 1.1c), the gas phase is modeled using the standard Eulerian approach described above and the spray is represented by a number of discrete computational 'particles'. Individual particles are tracked through the flow domain from their injection point until they escape the domain in a Lagrangian frame work (Nijdam et al. 2006). The Eulerian–Lagrangian model has the advantage of being computationally cheaper than the Eulerian–Eulerian method for a large range of particle sizes. It can also provide more details of the behaviour and residence times of individual particles and can potentially approximate mass and heat transfer more accurately. On the other hand, the approach can be expensive if a large number of particles have to be tracked and it is best when the dispersed phase does not exceed 10 % by volume of the mixture in any region (Marshall and Bakker 2002).

In both the Eulerian–Eulerian and the Eulerian–Lagrangian methods, the exchange of momentum between particles and gas needs to be modeled. This exchange can consist of several forces such as drag, lift, virtual mass, and wall forces. Mostafa and Mongia (1987) concluded that the Eulerian approach performs better than Lagrangian method. In contrast, Nijdam et al. (2006) found that both Eulerian and Lagrangian modeling approaches gave similar predictions for turbulent droplet dispersion and agglomeration of sprays for a wide range of droplet and gas flows. The two models were found to require similar computing times for a steady axi-symmetric spray. However, the authors preferred the Lagrangian models because of their wider range of applicability.

## 1.5  CFD Analysis

CFD analysis involves following three main steps. The first step is *pre-processing*, which includes problem definition, geometry, meshing (this can usually be done with the help of a standard CAD program), and generation of a computational model. The second step is *processing*, which uses a computer to solve the mathematical equations of fluid flow. The final step of *post-processing* is used to evaluate and visualize the data generated by the CFD analysis (Xia and Sun 2002) and validate

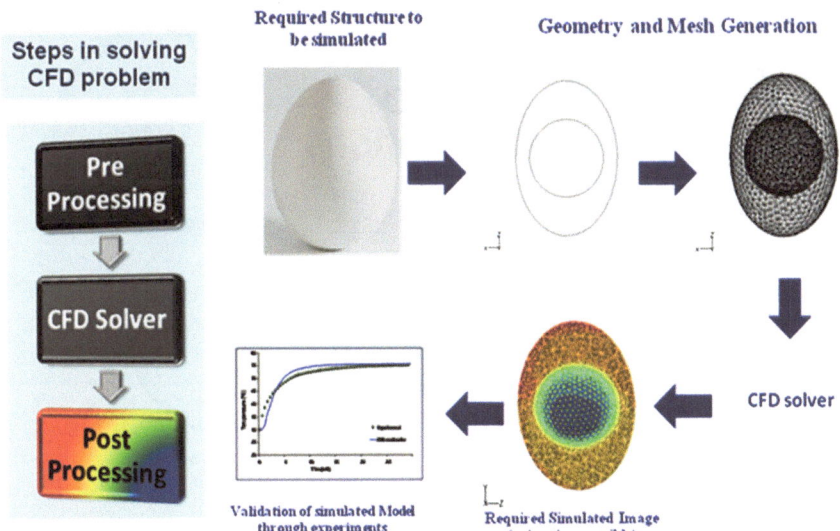

**Fig. 1.2**  CFD simulation steps

the simulation results with experimental data. Figure 1.2 explains the all three steps of CFD analysis with the example of egg pasteurization.

## 1.6 CFD Applications in Food Processing

Although the origins of CFD can be found in the automotive, aerospace and nuclear industries with a variety of applications in different processing industries, it is only in recent years that CFD has been applied to food processing (Scott and Richardson 1997). The applications of CFD in the food industry have been reviewed by many researchers (Scott and Richardson 1997; Xia and Sun 2002; Anandharamakrishnan 2003; Norton and Sun 2006). These reviews envisage the potential of CFD to be used as a tool in predicting the fluid flow, heat and mass transfer phenomena in the food processes, leading to better equipment design and process control for the food industry. Figure 1.3 depicts the applications of CFD in various food processing operations.

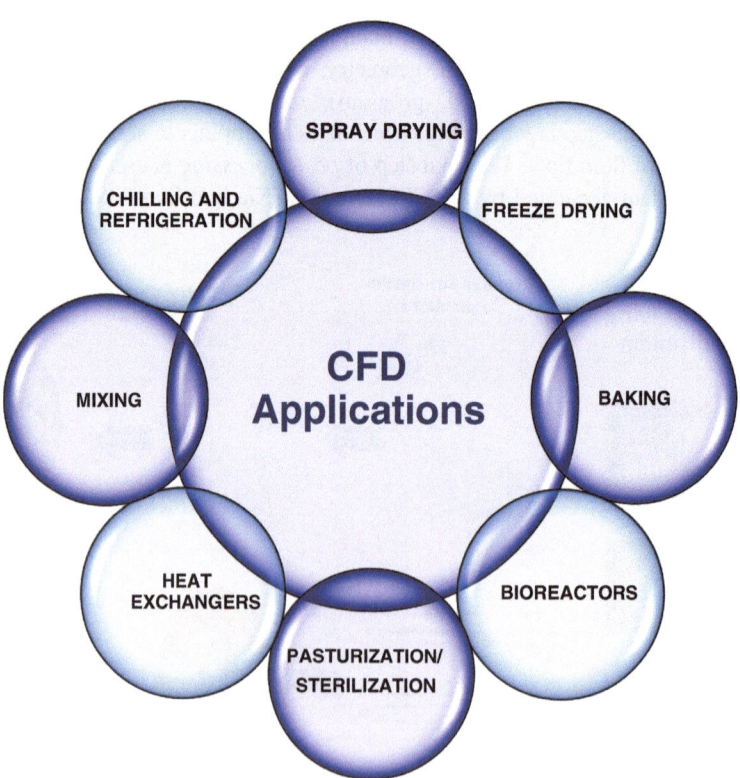

**Fig. 1.3** CFD applications in food processing

The main application of CFD includes spray drying processes (Langrish and Fletcher 2001, 2003), baking process (Therdthai et al. 2003; DeVries et al. 1994; Mills 1998–1999), refrigerated display cabinets (Cortella et al. 1998), thermal sterilization (Datta and Teixeira 1987; Abdul Ghani et al.1999a, b, 2001), pasteurization of egg (Denys et al. 2003, 2004, 2005), mixing (Sahu et al. 1999; Scott 1977), refrigeration (Hu and Sun 1999, 2000; Davey and Pham 1997, 2000; Moureh and Derens 2000; Mariotti et al. 1995), spray freezing (Anandharamakrishnan et al. 2010b), heating and cooling processes (Wang and Sun 2003), and humidification of cold storage (Verboven and Nicolai 2008, 2009). The list given above is non-exhaustive, and for detailed review of CFD applications to food processing, reader may refer elsewhere (Sun 2007). CFD has recently found widespread applications in food processing. In thermal sterilization processes, CFD has found increased use in analyzing the flow pattern, temperature distribution, and more importantly, the shape and position of the slowest heating zone (SHZ), since it is very difficult to estimate these parameters using experiments.

## 1.7  Nomenclature

| | |
|---|---|
| $C_{1\varepsilon}, C_{2\varepsilon}, C_\mu$ | Constants |
| $E$ | Internal (thermal) energy (J/mol) |
| $h$ | Enthalpy (J/kg) |
| $g$ | Gravitational force (m/s$^2$) |
| $G_k$ | Generation of kinetic energy |
| $\underline{J}$ | Diffusion flux (kg/m$^2$. s) |
| $k$ | Turbulence kinetic energy |
| $k_{eff}$ | Effective conductivity (W/mK) |
| $k_{ta}$ | Thermal conductivity (W/mK) |
| $k_t$ | Turbulent thermal conductivity (W/mK) |
| $p$ | Pressure (Pa) |
| $T$ | Temperature (K) |
| $t$ | Time (s) |
| $v$ | Velocity (m/s) |
| $\varepsilon$ | Turbulence dissipation rate (m$^2$/s$^3$) |
| $\sigma_k, \sigma_\varepsilon$ | Turbulent Prandtl numbers |
| $\rho$ | Density (kg/m$^3$) |
| $\rho_g$ | Gas density (kg/m$^3$) |
| $\tau$ | Stress tensor (N/m$^2$) |
| $\mu_t$ | Turbulent (or eddy) viscosity (kg/ms) |
| $\mu$ | Viscosity (kg/ms) |

# Chapter 2
# Computational Fluid Dynamics Applications in Spray Drying of Food Products

Spray drying is a well-established method for converting liquid feed materials into a dry powder form. It is widely used to produce powdered food, healthcare and pharmaceutical products. Normally, spray dryer comes at the end-point of the processing line, as it is an important step to control the final product quality. It has some advantages, such as rapid drying rates, a wide range of operating temperatures and short residence times. In spray drying operations, CFD simulation tools are now often used, because measurements of air flow, temperature, particle size and humidity within the drying chamber are very difficult and expensive to obtain in large-scale dryer (Kuriakose and Anandharamakrishnan 2010).

## 2.1 Spray Drying Process

Spray drying is the process of transforming a feed (solution or suspension) from a fluid into a dried particulate form by spraying the feed into a hot drying medium. Spray drying is a widely used industrial process for the continuous production of dry powders with low moisture content (Charm 1971; Masters 1991; Anandharamakrishnan et al. 2007). As shown in Fig. 2.1, spray drying involves four stages of operation: (1) atomization of liquid feed into a spray chamber; (2) contact between the spray and the drying medium; (3) moisture evaporation; and (4) separation of dried products from air stream.

### 2.1.1 Atomization

Atomization is a process where the bulk-liquid breaks up into a large number of small droplets. The choice of atomizer is most important in achieving economic production of high quality products (Fellows 1998). The different types of atomizer (Masters 1991) are:

C. Anandharamakrishnan, *Computational Fluid Dynamics Applications*
*in Food Processing*, SpringerBriefs in Food, Health, and Nutrition,
DOI: 10.1007/978-1-4614-7990-1_2, © Chinnaswamy Anandharamakrishnan 2013

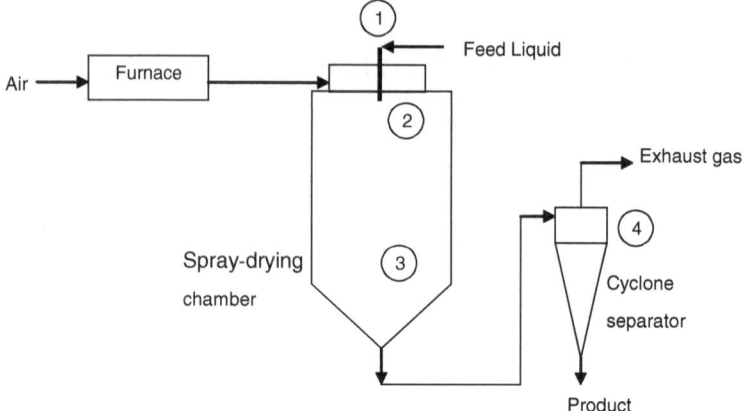

**Fig. 2.1** Processing stages of spray dryer

*Centrifugal or rotary atomizer*: Liquid is fed to the center of a rotating wheel with a peripheral velocity of 90–200 m/s. Droplets are produced typically in the range of 30–120 μm sizes. The size of droplets produced from the nozzle varies directly with feed rate and feed viscosity, and inversely with wheel speed and wheel diameter.

*Pressure nozzle atomizer*: Liquid is forced at 700–2000 kPa pressure through a small aperture. Here the size of droplets is typically in the range of 120–250 μm. The droplet size produced from the nozzle varies directly with feed rate and feed viscosity, and inversely with pressure.

*Two-fluid nozzle atomizer*: Compressed air creates a shear field, which atomizes the liquid and produces a wide range of droplet sizes.

## 2.1.2  Spray–Air Contact

During spray–air contact, droplets usually meet hot air in the spraying chamber either in co-current flow or counter-current flow. In co-current flow, the product and drying medium passes through the dryer in the same direction. In this arrangement, the atomized droplets entering the dryer are in contact with the hot inlet air, but their temperature is kept low due to a high rate of evaporation taking place, and is approximately at the wet-bulb temperature. As the droplets pass through the dryer, the moisture content decreases, the air temperature also decreases, and so the particle temperature does not rise substantially as the particle dries and the effect of evaporation cooling diminishes (Mujumdar 1987). The temperature of the products leaving the dryer is slightly lower than the exhaust air temperature. This co-current configuration is therefore very suitable for the drying of heat-sensitive materials. The advantages of the co-current flow process are rapid spray evaporation, shorter evaporation time and less thermal degradation of the products (Masters 1991; Anandharamakrishnan et al. 2007).

In contrast, in the counter-current configuration, the product and drying medium enter at the opposite ends of the drying chamber. Here, the outlet product temperature is higher than the exhaust air temperature, and is almost at the feed-air temperature with which it is in contact. This type of arrangement is used for non-heat sensitive products only. In another type called mixed flow, the dryer design incorporates both co-current flow and counter-current flow. This type of arrangement is used for drying of coarse free-flowing powder, but the drawback is that the temperature of the product is high (Masters 1991).

### 2.1.3  Moisture Evaporation

When droplets come in contact with hot air, evaporation of moisture from their surfaces takes place. The large surface area of the droplets leads to rapid evaporation rates, keeping the temperature of the droplets at the wet-bulb temperature (Mujumdar 1987). In this period, different products exhibit different characteristics, such as expansion, collapse, disintegration and irregular shape. Methods for calculating the changes in size, density and studies of droplet drying are described by Masters (1991).

### 2.1.4  Separation of Dried Products

The dry powder is collected at the base of the dryer and removed by a screw conveyor or a pneumatic system with a cyclone separator. Other methods for collecting the dry powder are bag filters and electrostatic precipitators (Fellows 1998). The selection of equipment depends on the operating conditions, such as particle size, shape, bulk density, and powder outlet position.

## 2.2  Types of Spray Dryers

The two main designs of commonly used spray dryers are the short-form and tall-form driers shown in Fig. 2.2.

Tall-form designs are characterized by height-to-diameter aspect ratios of greater than 5:1. Short-form dryers have height-to-diameter ratios of around 2:1. The short-form dryers are the most widely used, as they accommodate the comparatively flat spray disk from a rotary atomizer (Masters 1991). The flow patterns observed in short-form dryers are more complex than those in tall-form dryers, with many dryers having no plug-flow zone and a wide range of gas residence times (Langrish and Fletcher 2001).

**Fig. 2.2** Schematic diagrams of spray dryer (Langrish and Fletcher 2001)

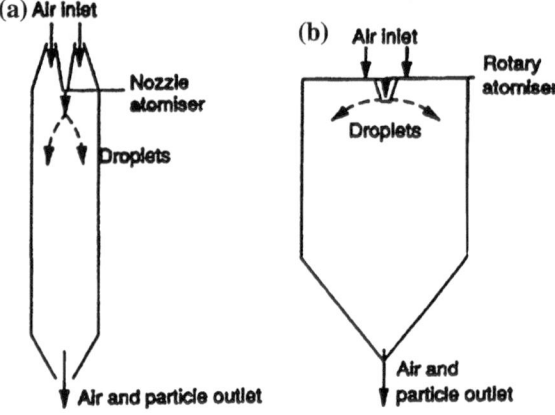

## 2.3  Airflow Pattern

During spray drying, the particle behaviour is dependent on the air flow pattern. Inside the spray chamber, there is presence of significant air flow instabilities due to the inlet swirl. The various spray drying–air flow studies have been summarized in Table 2.1. Hence, the effect of turbulence inside the spray chamber should be considered. Huang et al. (2004) showed that RNG k-ε model prediction was better for swirling two-phase flow in the spray drying chamber compared to standard k-ε, realizable k-ε and Reynolds stress models.

The air flow patterns in an industrial spray dryer used for milk powder production have been modeled using the transient Reynolds-averaged Navier–Stokes equations with the Shear Stress Transport (SST) turbulence model (Gabites et al. 2010). These simulations were carried out in the absence of atomized liquid droplets. The simulations showed that the main air jet oscillated and processed about the central axis with no apparent distinct frequency. In turn, the recirculation zones between the main jet and the chamber walls fluctuated in size. Good agreement was found between the movements of the main jet via simulations and from telltale tufts installed in the plant dryer. The different outlet boundary condition appeared to have little influence on the overall flow field. In the gas-only simulations, different fluid bed flows within the range had only a local influence by reducing the length of the main jet. This may have an effect on the particle capture by the fluid bed.

## 2.4  Atomization

The atomization stage during spray drying is very important, since it affects the final particle size. A co-current spray dryer fitted with pressure nozzle was investigated both in experiment and CFD simulation by Kieviet et al. (1996), to develop

**Table 2.1** Spray drying–airflow pattern studies (Kuriakose and Anandharamakrishnan 2010)

| Problem descriptions | Turbulence model | Findings | Authors |
|---|---|---|---|
| Simulation of airflow pattern with experimental validation. | Standard k-ε and RSM | Non-swirling flow spray chamber; the k-ε model gives good predictions of gas velocity profiles, whereas for swirling flows, RSM model gives better accurate predictions. | Oakley and Bahu (1993) |
| Simulation of airflow pattern to find out the oscillations in the flow field. | Standard k-ε | Strongest oscillations occur. Good agreement between hot-wire anemometer velocity measurements and simulation results. | Langrish et al. (1993) |
| Effects of the air inlet geometry and spray cone angle on wall deposition rates. | Standard k-ε | High swirl in the inlet air and large spray cone angle gave the lowest wall deposition rates in both the experiments and simulation. | Langrish and Zbicinski (1994) |
| Simulation of airflow and particle trajectories in the tall-form dryer with experimental validation. | Standard k-ε | Good agreement between measurements and simulation results. | Zbicinski (1995) |
| Simulation of airflow pattern, temperature, humidity, particle trajectories and resistance time in a co-current spray dryer fitted with a pressure nozzle. | Standard k-ε | Model prediction agreed well with the experimental measurements of velocity, temperature and humidity. | Kieviet (1997) |
| Simulation studies on the effects of increased turbulence in inlet airflow. | Standard k-ε | An increase in the amount of evaporation resulted directly from enhanced inlet turbulence. | Southwell et al. (1999) |
| Temperature and moisture content of the air with the trajectories of the particles. | Standard k-ε | The drying of droplets is influenced by particles surface to surrounding air and diffusion within the particles. | Straatsma et al. (1999) |
| Investigating the airflow pattern, temperature, velocity and humidity profile at different spray dryer chamber configuration. | Standard k-ε | The optimal chamber geometry will depends on the feed properties, atomizer type and drop size distribution | Huang et al. (2003b) |

(continued)

**Table 2.1** (continued)

| Problem descriptions | Turbulence model | Findings | Authors |
|---|---|---|---|
| Experimental and simulation studies of inlet air swirl on the stability of the flow pattern in spray dryers. | RSM | Comparison of with and without spray showed that the introduction of spray has significant effect on the flow behavior. An increase in swirl angle changes the internal flow pattern. | Langrish et al. (2004) |
| Simulation of a spray dryer with rotary atomizer. Kieviet's (1997) spray dryer geometry was used. | Standard k-ε, RNG k-ε, Realizable k-ε and RSM | Realizable k-ε cannot be used to simulate highly swirling two-phase flow. RNG k-ε turbulent model gives adequate accuracy at reasonable computational time. | Huang et al. (2004) |
| Simulation of spray dryer fitted with rotary atomizer. | RNG k-ε | More volume of drying chamber is used by rotary atomizer and existence of strong reverse flow just beneath the rotating disc. | Huang et al. (2005) |
| Simulation of a spray dryer with pressure nozzle and rotary atomizer. Kieviet's (1997) spray dryer geometry was used. | RNG k-ε | Simulation results agreed well with Kieviet (1997) experimental results. | Huang et al. (2006) |
| Simulation of a spray dryer with rotary atomizer | RANS | Rotary atomizer has a big influence on the flow pattern in pilot scale spray dryer, but its influence decreases with increase in size of spray dryer. | Ullum (2006) |
| Simulation of industrial scale spray dryer with a new drying kinetics model for a heat-sensitive solution. | Standard k-ε | Good agreement with experimental data. Off-design performance of spray dryer was predicted to analyze the effect of various operating parameters on drying performance. | Huang and Mujumdar (2007) |
| Evaluation of droplet drying models in a spray dryer fitted with rotary atomizer using CFD simulation | RNG k-ε | The concept of particle rigidity prediction in a CFD simulation was explored, and the effect of initial feed moisture content on the drying models was also studied. | Woo et al. (2008) |

(continued)

**Table 2.1** (continued)

| Problem descriptions | Turbulence model | Findings | Authors |
|---|---|---|---|
| Modeling droplet drying in a spray dryer fitted with a pressure nozzle under steady and unsteady state. | Standard k-ε | 2D models can be used for fast and low-resource-consumption numerical calculations with some drawbacks. 3D models can predict the asymmetric flow patterns and provide actual 3D picture of particle trajectories, but require high computing effort. | Mezhericher et al. (2009) |
| Simulation of industrial scale spray dryer attached with a fluidized bed, using Reaction Engineering Approach (REA). | Standard k-ε | Smaller spray cone angle facilitates easy movement of particles to the fluidized bed. The accuracy of REA model in predicting the single droplet drying kinetics was also explained. | Chen and Jin (2009a) |

a theoretical model for the prediction of final product quality. Good agreement was obtained between the experimental data and the simulation. An ultrasonic nozzle spray dryer was studied numerically by Huang et al. (2004). Birchal et al. (2006) simulated a spray dryer fitted with a rotary atomizer for drying of milk emulsion by using CFD, and also by a model with simplified particle motion. Authors also discussed the advantages and limitations of each model in the design and optimization of spray dryers. Studies on the effects of atomizer types (rotary disc and pressure nozzle) on droplet behaviour were performed by Huang et al. (2006) using CFD for spray drying of maltodextrin. They concluded that pressure nozzle may lead to a higher velocity variation in the center of the chamber than the rotary atomizer. Moreover, large recirculation of droplets was also found during pressure nozzle atomization.

## 2.5  Particle Histories

The understanding of particle histories, such as velocity, temperature, residence time and the particle impact position, are important to design and operate a spray dryer. Moreover, final product quality is dependent on these particle histories. These particle histories can be tracked with the help of CFD simulations.

## 2.6  Air–Particle Interaction

The primary problem in spray drying modeling is the coupling of equations in mass, momentum and energy between the gas and the droplets. These coupling phenomena of mass transfer from droplet to gas were coupling by evaporation, momentum exchange via drag, and energy coupling by heat transfer, which are schematically shown in Fig. 2.3.

Heat is transferred from the gas phase to the droplets convectively, and this leads to a decrease in temperature of the gas, which it affects the viscosity and density of the gas, which may in turn affect the gas flow field. This also affects the droplet trajectories

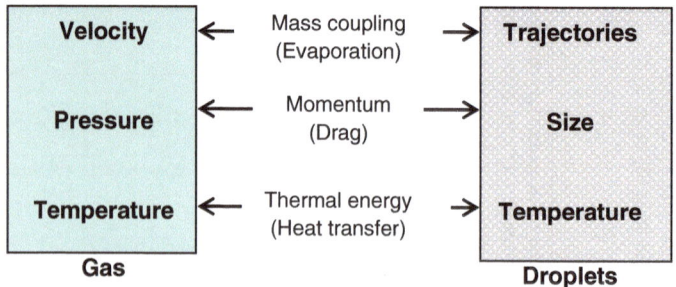

**Fig. 2.3**  Gas-droplet coupling phenomena

and the heat transfer rate between the droplets and the gas (Crowe et al. 1977). Hence, all three equations (mass, momentum and energy) are interdependent and should be included in the gas-droplet interactions (Kuriakose and Anandharamakrishnan 2010).

## 2.7 Particle Tracking

Both the Eulerian–Eulerian and Eulerian–Lagrangian methods have been used in published simulations of spray dryings. However, the Eulerian-Lagrangian frame work was selected most often, because it provides residence time of individual particles with a large range of particle sizes. Crowe et al. (1977) first proposed the particle source in the cell (PSI-Cell) model. This is the basis for the discrete phase model (DPM).

In the DPM, the flow field is divided into a grid defining computational cells around each grid point. Each computational cell is treated as a control volume for the continuous phase (gas phase). The droplets are treated as source of mass, momentum and energy inside the each control volume. The gas phase is regarded as a continuum (Eulerian approach), and is described by first solving the gas flow field, assuming no droplets are present. Using this continuous phase flow field, droplet trajectories, together with size and temperature histories along the trajectories, are calculated. The mass, momentum and energy source terms for each cell throughout the flow field is then determined. The source terms are evaluated from the droplet equation and are integrated over the time required to cross the length of the trajectory inside each control volume. The results are multiplied (scaled up) by the number flow rate of drops associated with this trajectory (Crowe et al.1977; Papadakis and King 1988; Fluent 2006). The gas flow field is solved again, incorporating these source terms, and then new droplet trajectories and temperature histories are calculated. This approach provides the influence of the droplets on the gas velocity and temperature fields. The method proceeds iteratively calculating gas and particle velocity fields.

The range of droplet sizes produced by the atomizer is represented by a number of discrete droplet sizes. Each initial droplet size is associated with one trajectory; along with the number of drops it is constant, assuming that no coalescence or shattering occurs. Once the air velocities, temperatures, and humidity are postulated, the transport equations for the droplets of each size are integrated over time and positioned to yield droplet trajectories, velocities, sizes and temperatures. Calculations for droplets of each initial size continue until the volatile fractions (e.g. water) in the droplets evaporate completely, exit the column, or impact the column wall (Papadakis and King 1988; Fluent 2006).

In the CFD simulation, a combined Eulerian and Lagrangian model is used to obtain particle trajectories by solving the force balance equation:

$$\frac{d\underline{u}_p}{dt} = \frac{18\mu}{\rho_p d_p^2} \frac{C_D Re}{24} \left(\underline{v} - \underline{u}_p\right) + \underline{g}\left[\frac{\rho_p - \rho_g}{\rho_p}\right] \qquad (2.1)$$

where $\underline{v}$ is the fluid phase velocity, $u_p$ is the particle velocity, $\rho_g$ is the density of the fluid and $\rho_p$ is the density of the particle.

The particle force balance (equation of motion) includes discrete phase inertia, aerodynamic drag and gravity. The slip Reynolds number ($Re$) and drag coefficient ($C_D$) are given in the following equations:

$$Re = \frac{\rho_g d_p \left| \underline{u}_p - \underline{v} \right|}{\mu} \tag{2.2}$$

$$C_D = a_1 + \frac{a_2}{Re} + \frac{a_3}{Re^2} \tag{2.3}$$

where $d_p$ is the particle diameter, and $a_1, a_2$ and $a_3$ are constants that apply to smooth spherical particles over several ranges of Reynolds number ($Re$) given by Morsi and Alexander (1972).

The velocity of particles relative to air velocity was used in the trajectory calculations (Eq. 2.1). Turbulent particle dispersion was included in this model as discrete eddy concept (Langrish and Zbicinski 1994). In this approach, the turbulent air flow pattern is assumed to be made up of a collection of randomly directed eddies, each with its own lifetime and size. Particles are injected into the flow domain at the nozzle point, and envisaged to pass through these random eddies until they impact the wall or leave the flow domain through the product outlet.

The heat and mass transfer between the particles and the hot gas is derived following the motion of the particles:

$$m_p c_p \frac{dT_p}{dt} = h A_p \left( T_g - T_p \right) + \frac{dm_p}{dt} h_{fg} \tag{2.4}$$

where $m_p$ is the mass of the particle, $c_p$ is the particle heat capacity, $T_p$ is the particle temperature, $h_{fg}$ is the latent heat, $A_p$ is the surface area of the particle, and $h$ is the heat transfer co-efficient.

The heat transfer coefficient ($h$) is obtained from the Ranz-Marshall equation.

$$Nu = \frac{h d_p}{k_{ta}} = 2 + 0.6 \left( Re_d \right)^{1/2} \left( Pr \right)^{1/3} \tag{2.5}$$

where Prandtl number ($Pr$) is defined as follows

$$Pr = \frac{c_p \mu}{k_{ta}} \tag{2.6}$$

where $d_p$ is the particle diameter, $k_{ta}$ is the thermal conductivity of the fluid, $\mu$ is the molecular viscosity of the fluid.

The mass transfer rate (for evaporation) between the gas and the particles is calculated from the following equation:

$$\frac{dm_p}{dt} = -k_c A_p \left( Y_s^* - Y_g \right) \tag{2.7}$$

where $Y_s^*$ is the saturation humidity, $Y_g$ is the gas humidity, and $k_c$ is the mass transfer co-efficient and can be obtained from Sherwood number:

$$Sh = \frac{k_c d_p}{D_{i,m}} = 2 + 0.6 \, (Re_d)^{1/2} \, (Sc)^{1/3} \tag{2.8}$$

where $D_{i,m}$ is the diffusion coefficient of water vapour in the gas phase and $Sc$ is the Schmidt number, defined as follows:

$$Sc = \frac{\mu}{\rho_g D_{i,m}} \tag{2.9}$$

The values of vapour pressure, density, specific heat and diffusion coefficients can be obtained from Perry (1984).

When the temperatures of the droplet has reached the boiling point and the mass of the droplet exceeds the non-volatile fraction, then the boiling rate model is applied (Kuo 1986).

$$\frac{d \, (d_p)}{dt} = \frac{4 k_{ta}}{\rho_p c_g d_p} \left( 1 + 0.23 \sqrt{Re} \right) \ln \left[ 1 + \frac{c_g \, (T_g - T_p)}{h_{fg}} \right] \tag{2.10}$$

where $k_{ta}$ is the thermal conductivity of the gas and $c_g$ is the heat capacity of the gas (Kuriakose and Anandharamakrishnan 2010).

## 2.8 Particle Temperature

The particle temperature is very important in the case of heat sensitive products, since it influences the aroma retention and thermal stability of heat labile components. Crowe et al. (1977) predicted that the smaller size particles have higher temperatures than the larger particles, because the latter have a smaller surface area to volume ratio and evaporate more slowly. Kieviet (1997) studied the airflow pattern, temperature, humidity, particle trajectories and residence time in a 2D co-current spray dryer fitted with a pressure nozzle using maltodextrin as feed solution, and concluded that the gradients in the center region of the drying chamber could be improved. Anandharamakrishnan et al. (2010a) studied the particle temperature in both short-form and tall-form spray dryer using CFD simulation for drying of whey proteins. They found that due to moisture evaporation of droplets, the temperature of droplets was high and was almost equal to the gas temperatures outside the core region. Moreover, the temperature of gas in the core spray region and the upper part of the chamber decreased due to the cooling effects of evaporation. The particle nature was also affected by the outlet air temperature (Kuriakose and Anandharamakrishnan 2010).

## 2.9 Residence Time of Particle

The particle residence time has a great impact on the final powder quality and it also affects product qualities such as solubility and bulk density. The residence time (RT) is divided into two parts; namely, primary and secondary residence times.

The primary RT is calculated from the time taken for droplets leaving the nozzle to impact on the wall or leave at the outlet. The secondary residence time can be defined as the time taken for a particle to slide along the wall from the impact position to the exit (Kuriakose and Anandharamakrishnan 2010).

Kieviet and Kerkhof (1995) determined the RTD of particles in a co-current spray dryer during the drying of aqueous maltodextrin solutions. Kieviet (1997) observed that during spray drying of maltodextrin solution, the larger diameter particles have longer RTs than smaller particles. He also found enormous difference between measured and predicted results due to particle wall depositions and sliding movement. Ducept et al. (2002) performed an experiment to determine the RTD of particles, and validated with the CFD predictions in a superheated steam spray dryer. The residence time distribution of different sized particles in a spray dryer was studied by Huang et al. (2003a), and they found that different droplets follow different trajectories in the drying chamber.

Anandharamakrishnan et al. (2010a) studied Particle Residence Time Distribution of whey proteins in both short-form and tall-form dryers and the residence time (Fig. 2.4). The study indicates that most of the particles have very low RT during spray drying (short-form). It was observed that a bent outlet pipe inside the chamber increases gas and particle recirculation (Fig. 2.4); consequently, cold gas is mixed with down-flowing hot inlet gas, and dried particles will be exposed to the high inlet gas

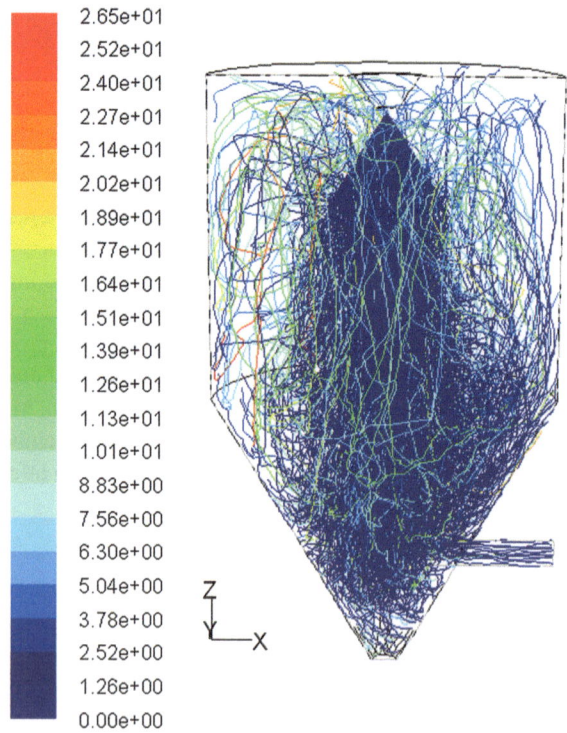

**Fig. 2.4** Particle trajectories colored by residence time(s) (Anandharamakrishnan et al. 2010a)

2.65e+01
2.52e+01
2.40e+01
2.27e+01
2.14e+01
2.02e+01
1.89e+01
1.77e+01
1.64e+01
1.51e+01
1.39e+01
1.26e+01
1.13e+01
1.01e+01
8.83e+00
7.56e+00
6.30e+00
5.04e+00
3.78e+00
2.52e+00
1.26e+00
0.00e+00

temperatures. This recirculation may lead to denaturation of proteins or inactivation of enzymes. Hence, bend outlet pipe needs to be avoided inside the chamber for producing high quality spray dried food products. Moreover, they found a large difference between the gas and particle residence time. However, there is no direct measurement of primary RT available to confirm the predictions, and this is an interesting challenge for future research (Kuriakose and Anandharamakrishnan 2010).

## 2.10  Particle Deposition and Position

The knowledge of particle impact positions is important for the design and operation of spray dryers, as it influences the final product quality. In an earlier numerical study, Reay (1988) has shown that the most likely areas for wall deposition are an annular area of the dryer roof and a region below the atomizer, where large particles are likely to deposit. Later, Kieviet (1997) investigated the interaction of wall deposition with the residence time, and the effect of wall deposition on the product quality and yield during spray drying of maltodextrin. Goula and Adamopoulos (2004) determined the operating conditions that influence the fouling and residue accumulation of the equipment during the drying process. Anandharamakrishnan et al. (2010a) studied the particle impact position during drying of whey proteins from the simulation data using an in-house post-processor. Figure 2.5a, b shows the top and front cross-sectional views of the simulated results (Anandharamakrishnan et al. 2010a). Figure 2.5a, b indicates that a large fraction of the particles (50 %) strike the conical

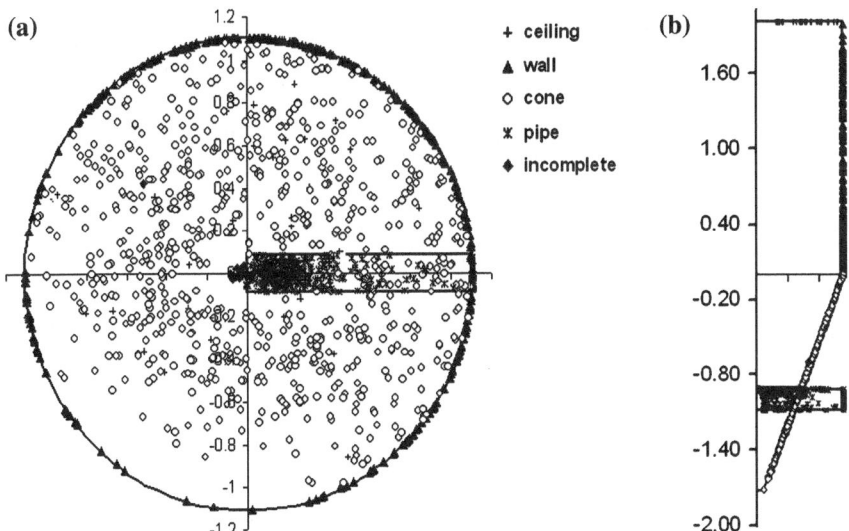

**Fig. 2.5**  Particle impact positions, **a** top view, **b** front view (Anandharamakrishnan et al. 2010a)

part of the spray dryer chamber (similar with the earlier observation of Langrish and Zbicinski 1994) and 23 % of particles hit the cylindrical part of the wall, but only a small proportion (25 %) of the particles come out of the outlet pipe line (the intended destination). A very small 2 % of particles hit the ceiling despite the large volume of re-circulated gas, but particles hitting the cone and/or cylindrical wall (73 %) should slide down to the main outlet aided by mechanical hammer operations. They also found that in a short-form dryer, a large fraction of the particles strike the conical part of spray dryer chamber, while in tall-form dryer, the particles struck the cylindrical part of the wall. In both forms of dryer, they found less impact on the ceiling, despite the recirculation of gas in the zone (Kuriakose and Anandharamakrishnan 2010).

## 2.11  Current Trends

In recent years, application of the Reaction Engineering Approach (REA), drying kinetics model, droplet–droplet interactions, unsteady state modeling and population balance model for the simulation of spray dryers has been increasing. The Reaction Engineering Approach assumes that evaporation is an activation process to overcome an energy barrier, while this is not the case for condensation or adsorption. The basic concept of REA was described by Chen and Xie (1997) and Chen et al. (2001). This method describes the droplet drying trend, giving a detailed account of the temperature changes that occur within the droplet during the drying period; some experimental data are required to determine the model parameters. The REA model was used by Chen and Xie (1997) for the simulation of drying of thin-layer food materials such as kiwifruit, silica gel, potato and apple slices. Moreover, Huang et al. (2004) found that this approach (REA) fits in well with the fluent commercial CFD code for spray drying.

The experimental determination of spray drying kinetics was performed by Zbicinski et al. (2002). They determined the spray drying kinetics as a function of atomization ratio and drying agent temperature. They also proved that, based on the critical moisture content of the material, spray drying kinetics can be determined from the generalized drying curves. These lab-scale details can be used for scaling up the spray drying process. Further, Woo et al. (2008) analyzed the effect of wall surface properties on the deposition problem during spray drying using different drying kinetics. They concluded that proper selection of dryer wall material will provide potential alternatives for reducing the deposition problem. Roustapour et al. (2009) performed a CFD study for the drying of lime juice. They determined the drying kinetics based on experimental results of moisture content variation along the length of chamber, and numerically estimated residence time of droplets. The authors found that an increase in initial droplet diameter resulted in a lower particle residence time. CFD was used to gain more insights into the drying characteristics of the mono-dispersed droplets produced using a low velocity spray tower. Introduction of droplet and mass transfer did not significantly alter the flow field. Analysis revealed that the wet bulb region was significant in this tower.

Varying the inlet air temperature from 100 to 180 °C resulted in contrasting drying histories. These drying kinetics were then extended to assess the in situ crystallization phenomenon. For this spray drying tower, it was found that lower inlet temperature conditions favored a higher degree of crystallinity.

Droplet–droplet interactions during the spray drying were performed by applying the transient mode of calculations (Mezhericher et al. 2008). The droplet collisions influenced the temperature and humidity patterns, while their effect on velocity was less marked. They investigated both insulated and non-insulated spray chambers and reported that the insulation of a spray chamber will affect the airflow patterns, thereby affecting the droplet trajectories.

The modeling of spray dryers using the population balance method is gaining importance as the model accounts for droplet growth, coalescence and break up during the spray drying process. Nijdam et al. (2004) modeled the particle agglomeration within the spray chamber using two different frameworks, namely, Lagrangian and Eulerian. They validated their prediction using phase doppler anemometry (PDA) measurement, and found that in terms of ease of implementation and range of applicability, the Lagrangian approach is more suitable for modeling of agglomeration of particles. The modeling of droplet drying in the spray drying chamber by applying the unsteady mode of calculations (Mezhericher et al. 2009) showed that among 2D and 3D analyses, the latter predicts asymmetry of flow patterns in the spray chamber. Chen and Jin (2009b) performed transient 3D simulations in an industrial-scale spray dryer (15 m tall and 10 m wide). They observed that the particles make the central jet oscillate more non-linearly and that the frequency of oscillation decreases with increasing feed rate. Woo et al. (2009) have performed unsteady state simulations of spray drying and investigated the effect of chamber aspect ratio and operating conditions on flow stability. The authors observed that a large expansion ratio produces a more stable flow due to the limitations of jet fluctuations by outer geometry constriction.

## 2.12   Scope for Future Research

There remains scope for future research in the area of optimization of the spray drying process. Further work is needed to refine the turbulence models for the Lagrangian approach, in order to account for the various particle turbulence phenomena and particle–particle correlations. Modeling of particle agglomeration (including gas–particle interaction and particle–particle correlations), wall deposition (including nature of the product) and predicting particle residence time (including sliding movement of particles in the secondary residence time) during spray drying of food products is currently lacking. Hence, there is also scope for further study in the area to overcome problems like agglomeration, wall deposition, particle residence time, thermal degradation of particles and aroma loss (Kuriakose and Anandharamakrishnan 2010). Langrish (2007) has reported the same. Thus, the modeling approach may lead to better productivity and high-quality food products.

# Chapter 3
# Applications of Computational Fluid Dynamics in the Thermal Processing of Canned Foods

Though several food processing technologies have been developed with the aim of increasing the shelf-life of foods, thermal processing remains the most widely used food preservation technique. Thermal processing of canned foods can be divided into two major process methods: *in-container sterilization* and *in-flow sterilization* (Weng 2006). The food is usually packed in metal containers, glass bottles, retortable pouches, retortable cartons, etc. A variety of foods, including fruits, vegetables, meat, poultry, fish and dairy products, are being preserved by this method. In-flow sterilization process refers to the aseptic processing technique wherein the food products (mostly liquids) are sterilized prior to packaging (e.g. milk and fruit juices). This chapter provides insight into the applications of CFD in the thermal processing of canned foods; analyzing the liquid flow pattern; temperature and velocity profiles; and shape, size and position of the slowest heating zone (SHZ) and the associated biochemical changes in various types of canned foods.

## 3.1 Canning of Foods

Canning is the process of sealing foodstuffs hermetically in containers (tin cans or glass containers) and processing them by heat so as to store them for longer periods of time (Weng 2006). The high water content in fruits, vegetables, dairy products, etc. make them highly perishable when stored under normal conditions. Food spoilage mainly occurs due to the activity of enzymes present in food; oxidation of food constituents; moisture loss; and growth of microorganisms like bacteria, yeast and molds, etc. Moreover, in canning, the main concern is to prevent the growth of the heat-resistant bacterium *Clostridium botulinum*, which produces a lethal toxin. The processing temperature depends upon the acidity of the foods. Low acid foods (pH > 4.6) require processing at a temperature of about 121 °C, while acid foods (pH < 4.6) such as fruits need to be processed in boiling water at 100 °C.

C. Anandharamakrishnan, *Computational Fluid Dynamics Applications*
*in Food Processing*, SpringerBriefs in Food, Health, and Nutrition,
DOI: 10.1007/978-1-4614-7990-1_3, © Chinnaswamy Anandharamakrishnan 2013

## 3.2 Canned Solid–Liquid Food Mixtures

Recently, CFD modeling studies have been extended to the sterilization of canned solid–liquid food mixtures. An overview of the research work carried out in CFD applications in thermal processing of canned foods is depicted in Table 3.1.

The presence of solids was found to influence the velocity profile and the position of SHZ inside the can. Unlike liquid foods, solid–liquid food mixtures are assumed to be heated by both conduction and convection (Ghani and Farid 2006; Kiziltas et al. 2010; Rabiey et al. 2007). In the case of pineapple slices canned in liquid sucrose solution, the liquid was assumed to be heated by natural convection exhibiting the recirculation phenomenon (Ghani and Farid 2006), and the solid food (pineapple slices) were presumed to be heated by conduction. For the solid–liquid mixtures, solids are randomly dispersed in the liquid phase (canned peas); Kiziltas et al. (2010) observed slight changes in the velocity profile. This was attributed to the heat exchange and surface deflections while the flow is slowly moving through the stack of solid particles. The SHZ of canned solid–liquid food varied in comparison to canned liquid sterilization. Such an effect was expected, as the presence of the solid reduces the effect of natural convection in the fluid, causing the SHZ to move slightly upwards. On the other hand, in the case of asparagus canned in brine, Dimou and Yanniotis (2011) found that the shape of the solids and the available space between the asparagus does not restrict the flow of the brine, and thus, the SHZ was found to lie toward the bottom, at a height of 13.5 % of the can height from the bottom. A similar result was observed by Kiziltas et al. (2010) for canned peas in water, where the SHZ was found toward the bottom of the can. Furthermore, the sterilization of solid–liquid food mixtures (beef fat in water) under a high-pressure processing unit was studied by Ghani and Farid (2007). A forced convection phenomenon was observed in the liquid component. Solids foods are found to be heated more than the liquid, due to the difference in their compression heating coefficient.

## 3.3 Bacterial Deactivation Kinetics

The following $F_0$ value equation is used to quantify the effects of heat treatment and time with respect to the survival of a microorganism (Holdsworth and Simpson 2007).

$$F = \int 10^{(T-T_{\text{ref}})/Z} dt \qquad (3.1)$$

In the canning process, the $F_0$ value is used for the low-acid canned foods and refers to the sterilization value with a $z$ value of 10 K (Holdsworth and Simpson 2007; Weng 2006).

**Table 3.1** Overview of the research work carried out in CFD applications in thermal processing of canned foods

| Problem description | Findings | Authors |
|---|---|---|
| Numerical simulation of natural convection heating of canned, thick, viscous liquid food products | The bottom of the can heated up at a slower rate than that predicted by conduction heating. The slowest heating zone (SHZ) was located at the bottom center of the can | Kumar et al. (1990) |
| Transient temperature and velocity profiles in canned non-Newtonian liquid food during sterilization in a still-cook retort | The computed particle path predicted that the liquid just below the top center is exposed to minimum heat treatment, and should be of concern during the thermal process calculations | Kumar and Bhattacharya (1991) |
| Simulation of natural convection heating of canned food using CFD | Recirculating flow was observed inside the can. SHZ kept moving, and eventually stayed at a height of 10–15 % of the can height from the bottom | Ghani et al. (1999a) |
| Bacterial deactivation of canned liquid food during sterilization using CFD | The concentration of the live bacteria depends on both temperature distribution and flow pattern | Ghani et al. (1999b) |
| Thermal sterilization of canned foods in a 3D pouch using CFD | The SHZ was found to migrate at 30–40 % of the pouch height from the bottom | Ghani et al. (2001) |
| CFD analysis of transient temperature and velocity profiles in a horizontal can during sterilization | For horizontal cans, the SHZ moved downwards and eventually stayed at a height of 20–25 % of the can height from the bottom. However, the vertical cans showed faster heating, due to enhanced natural convection by its greater can height | Ghani et al. (2002a) |
| Theoretical and experimental investigation of the thermal inactivation of *Bacillus stearothermophilus* in food pouches | The measured concentration of *Bacillus stearothermophilus* spores at different periods of heating showed good agreement with the CFD simulation results. | Ghani et al. (2002b) |
| Theoretical and experimental investigation of the thermal destruction of Vitamin C in food pouches | SHZ migrated to 30–40 % of the pouch height. The vitamin profile depends on temperature and velocity profiles in the pouch. Simulation results agreed with the measured values | Ghani et al. (2002c) |

(continued)

**Table 3.1** (continued)

| Problem description | Findings | Authors |
|---|---|---|
| Heat transfer to a canned corn starch dispersion under intermittent agitation | Better heat transfer from the hot wall to the inner fluid was observed, since the intermittent rotation dislocated the gelatinized starch layer from the boundary region. | Tattiyakul et al. (2002) |
| The effect of can rotation on sterilization of liquid foods using CFD | The volume of SHZ covered only 5 % of the can volume at the end of heating. High fluid velocities were found close to the two ends of the horizontal can and gradually spread throughout the can length | Ghani et al. (2003) |
| A new computation technique for the estimation of sterilization time in canned food | A generalized correlation was developed to predict the change of SHZ over time for different fluids and can sizes | Farid and Ghani (2004) |
| Enhanced food sterilization through inclination of container walls and geometry modifications | Vertically oriented upright full conical cans gave the lowest sterilization time. Cylinders gave lower sterilization time when oriented horizontally compared to vertically. | Varma and Kannan (2005) |
| CFD modeling used to analyze the thermal sterilization of solid–liquid food mixture in cans | SHZ stayed at 30–35 % of the can height from bottom. The configuration of solids (pineapple slices) in the can significantly influence the rate of heating | Ghani and Farid (2006) |
| CFD studies on natural convective heating of canned food in conical and cylindrical containers | The SHZ temperature attained final fluid sterilization temperature the fastest for an upright conical geometry; this was followed by the cylinder and the downward pointing cone | Varma and Kannan (2006) |
| Simulation of sterilization of canned liquid food using sucrose degradation as an indicator | At 120 °C, $F_o$ showed a higher value than that obtained by the degradation of spores | Siriwattanayotin et al. (2006) |
| Numerical simulation of solid–liquid food mixture in a high pressure processing unit using CFD | The solids were found to be heated more than the liquid, due to the difference in their compression heating coefficient | Ghani and Farid (2007) |

(continued)

**Table 3.1** (continued)

| Problem description | Findings | Authors |
| --- | --- | --- |
| 3D simulations of heat transfer and liquid flow during sterilization of large particles in a cylindrical vertical can | The liquid flows upwards in a thin boundary layer and flows downwards in the interstice between the particles. SHZ occurs at the bottom of the can due to thermal stratification | Rabiey et al. (2007) |
| Heat transfer analysis of canned food sterilization in a still retort | Correlations were developed for Nusselt number as functions of Fourier number, food can aspect ratio, and thermal conductivity of the food medium | Kannan and Sandaka (2008) |
| CFD studies on pasteurisation of canned milk | Rotation of can with 5 rpm was more effective than stationary can during pasteurization | Anandpaul et al. (2011) |
| Simulation of heat transfer for solid–liquid food mixtures in cans | CFD simulation of the pasteurization of canned peas in water was validated with experimental measurements | Kiziltas et al. (2010) |
| Effect of can orientation on beer pasteurization evaluated using CFD | The package orientation did not result in process improvement | Augusto et al. (2010) |
| Velocity and temperature field characteristics of water and air during natural convection heating in cans | In the presence of headspace in cans, increased heating rates were observed. Organized velocity motions along the air–water interface resulted in vortex evolution | Erdogdu and Tutar (2010) |
| Evaluation of geometric symmetry condition of thermal process of packed liquid food using CFD modeling. | The axial symmetry of the bottles allows the use of smaller models, which saves computational effort | Augusto and Cristianini (2010a) |
| CFD analysis of viscosity influence on Thermal In-Package liquid food process | The sterilization process values became lower when the viscosity was increased up to a critical value | Augusto and Cristianini (2010b) |
| Investigation of asparagus sterilization process in a still can, using CFD modeling | Heating of the asparagus was not uniform. The top of the asparagus receives higher heat treatment than the bottom, due to faster heating of the top of the can | Dimou and Yanniotis (2011) |
| Quantifying enhancement in heat transfer due to natural convection during canned food thermal sterilization in a still retort | The transition to convection-augmented heat transfer from that of solely conduction mode occurred at a critical Fourier number | Koribilli et al. (2011) |

## 3.4 Analysis of Fluid Flow Pattern During the Thermal Sterilization Process

During the natural convection heating of canned liquid food products, the fluid movement accelerates the sterilization process by enhancing the rate of heat transfer. Hence, heat transfer by convection is governed by both fluid motion and temperature difference (Rao and Anantheswaran 1988). A common application of CFD and numerical simulation is the prediction of flow patterns and temperatures during the thermal processing of foods (Scott and Richardson 1997). During natural convection heating, the velocity in the momentum equations is coupled with temperature in the energy equation, because the movement of fluid is solely due to buoyancy force.

In the thermal sterilization of canned liquid foods, initially the liquid near the wall is at rest, due to the application of no-slip boundary condition. Later on, as the steam comes in contact with the outer walls of the can, the liquid adjacent to the wall heats up to the wall temperature by conduction. However, liquid away from the wall is still at the initial temperature. Because of gravity and the variation of liquid density throughout the can, buoyancy forces are created. Throughout the heating period, the buoyancy forces are opposed by the liquid viscous forces. The velocity of the convective current is dependent upon the strength of the buoyancy forces, and the magnitude of the liquid viscosity's resistance to flow. As the heating progresses, temperature-dependent viscosity decreases, resulting in an increased velocity. This results in faster heating of the product. Later on, as the temperature inside the container becomes uniform, the buoyancy forces decrease, leading to a significant reduction in velocity and cessation of recirculation (Kumar and Bhattacharya 1991; Ghani et al. 2001). For canned milk, the conduction pattern, followed by convection of temperature and velocity, has been observed with progression of time, as shown in Fig. 3.1 (Anandpaul et al. 2011).

## 3.5 Thermal Processing of Canned Fruits

The thermal processing of canned fruits is an important preservation technique used to improve the shelf-life of canned food products, through the inactivation of heat-resistant microorganisms and enzymatic activity that cause spoilage. CFD was used to analyze heat transfer and liquid flow during the sterilization of large particles in a cylindrical vertical can (Rabiey et al. 2007); heat transfer in canned peas under pasteurization conditions (Kiziltas et al. 2010); and thermal sterilization of canned solid–liquid food mixture (pineapple slices with its moisture) in cans (Ghani and Farid 2006). Recently, Padmavathi and Anandharamakrishnan (2012) analyzed the temperature profile during thermal processing of canned pineapple slices and pineapple tidbits using temperature-dependent thermo-physical properties, and validation of CFD simulation predictions with experimental results.

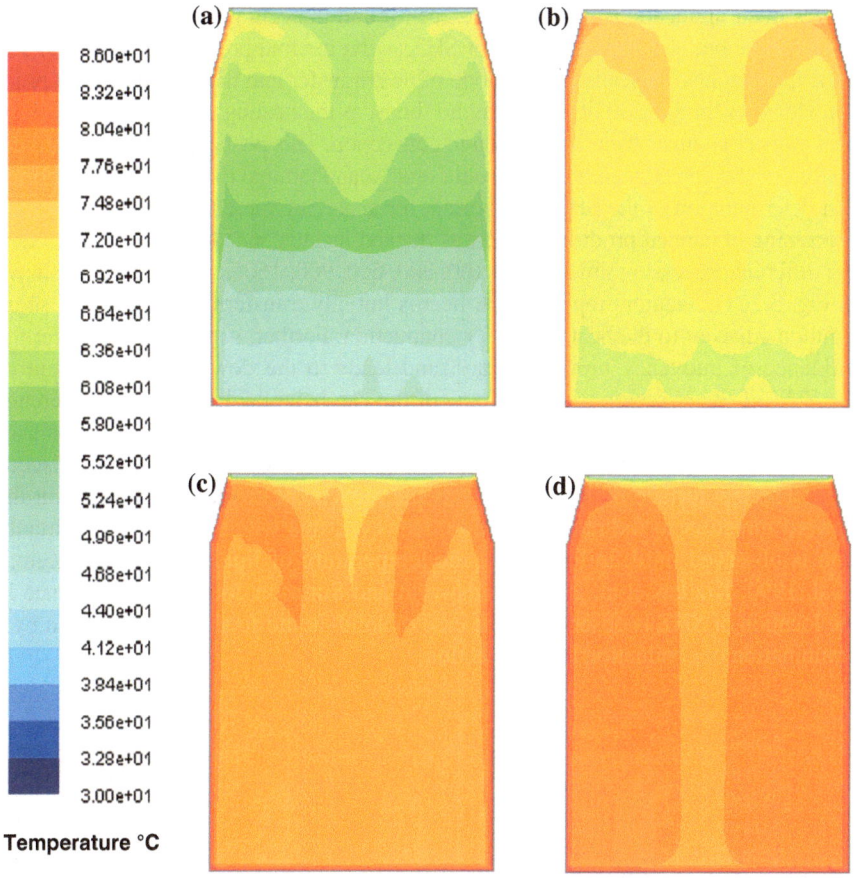

**Fig. 3.1** Temperature (°C) profiles of the pasteurization (85 °C) process of milk in stationary can position at **a** 60 s, **b** 120 s, **c** 240 s, and **d** 360 s (Anandpaul et al. 2011)

### 3.5.1 Temperature Profile and the Slowest Heating Zone

Temperature profile is the principal criteria for analyzing the effectiveness of a sterilization process. It is important to study the temperature distribution inside the can during the sterilization process. SHZ refers to a core region in the can that takes the longest time to reach the final sterilization temperature relative to the other regions, and hence represents the rate limitation (Varma and Kannan 2006). Therefore, the efficiency of a sterilization process is established by the temperature recorded at the SHZ, at the end of the sterilization process. The observation of the SHZ is a difficult task and requires detailed knowledge about transient flow patterns and temperature profiles (Ghani et al. 1999a). Furthermore, the SHZ do not have a constant location in the enclosure and changes with time, as well as with the geometry of the enclosure (Varma and Kannan 2006).

Widespread applications of CFD have been found in accurately predicting not only
the location, but also the movement of SHZ as the thermal processing proceeds. The
SHZ location also depends on the mode of heat transfer and the nature of food product. However, in the case of canned liquid foods, as the heating proceeds, the mode of
heat transfer changes from conduction to convection. This pushes the SHZ toward the
bottom of the can. Recently, Padmavathi and Anandharamakrishnan (2012) reported
that determination of the SHZ is important for analyzing the effectiveness of thermal
processing of canned products. SHZ was tracked inside canned pineapple slices during different processing time periods (60, 300, 600, 900, 1500, and 2100 s), as shown
in Fig. 3.2. The authors reported that heat is initially transferred to the liquid sugar
solution adjacent to the heated wall by conduction. Further, sugar solution under the
influence of buoyancy moves upward, and leads to the downward movement of
the SHZ of liquid region to about 10 % of the can height (Fig. 3.2). The difference
in the temperature due to natural convection creates a density gradient that makes
the hot sugar solution move upwards, and also carries the colder fluid by a viscous
drag. The top surface of the can deflects the movement of the fluid flowing in the
upward direction to the radial direction. However, this whole process of recirculation is observed until uniformity of the temperature of fluid is achieved (Kumar
et al. 1990; Kiziltas et al. 2010; Holdsworth and Simpson 2007). The difference in
the location of SHZ is observed in the case of solid–liquid food mixtures heated by a
combination of conduction and convection.

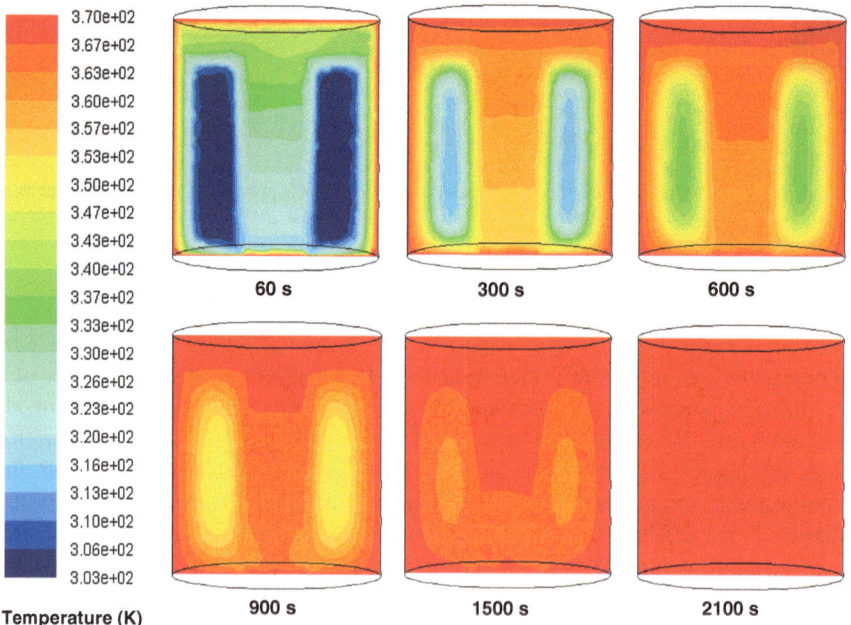

**Fig. 3.2** Temperature contours (*K*) of canned pineapple slices in sugar solution (Padmavathi and
Anandharamakrishnan 2012)

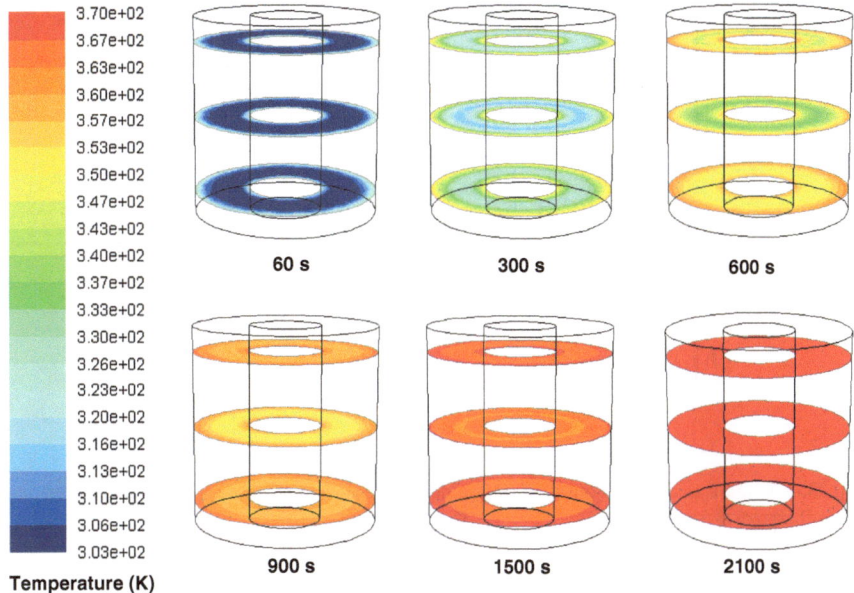

**Fig. 3.3** Temperature profiles of pineapple slices (Padmavathi and Anandharamakrishnan 2012)

In this regard, the SHZ of the canned pineapple slices is located radially away from the can's center, as shown in Fig. 3.2. Figure 3.3 shows the temperature profile of the pineapple slices during thermal processing at three zones: (1) center; (2) top, at a distance of 3.5 cm from the center; and (3) bottom, at a distance of 3.5 cm from the center. It clearly shows the sequence of the region getting heated. The heating of solid–liquid processes was often based on the assumption that the solid–liquid heats up by pure conduction, ignoring the effect of natural convection and leading to much longer than needed heating times, which results in over-processed products. Additionally, the effect of natural convection on the heating process of pineapple slices in a model liquid was verified by Ghani and Farid (2006), indicating that natural convection in the liquid has significant effects in the evolution of temperature and heat transfer rate.

### 3.5.2 $F_0$ Value During Thermal Processing of Canned Pineapple Slices

Microbial inactivation was shown to be in exponential relation with temperature, after the invasion of microbial inactivation, chemical and biochemical reactions. Considering this, thermal profiles become an important factor for evaluating the thermal process, resulting in sterilization values ($F_0$) (Augusto and Cristianini 2011). Hence, the effectiveness of a heating process was estimated based on $F_0$ values (Eq. 3.1).

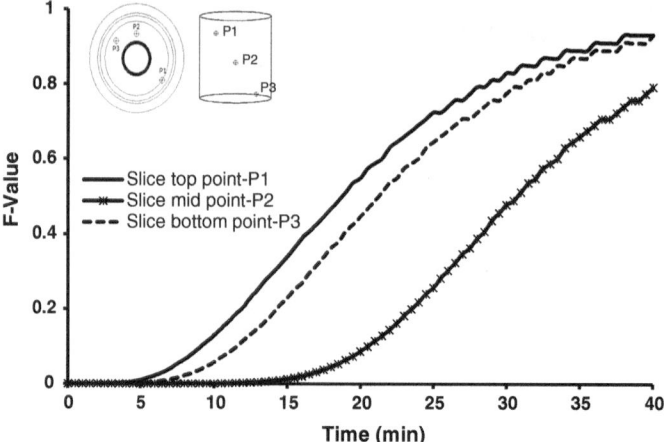

**Fig. 3.4** Graphical representation of $F_0$ value for canned pineapple slices (Padmavathi and Anandharamakrishnan 2012)

The survival of the microorganism can be quantified from process value $F_0$ (in minutes). The $F_0$ values were calculated for the canned pineapple slices at three different positions (Fig. 3.4) for every 1 min interval of the thermal process. Figure 3.4 shows that the pineapple slices attained a maximum $F_0$ value (higher inactivation) at 40 min of heating. Further studies in the area of thermal processing of canned solid–liquid foods are recommended, as the literature reports on such works are limited.

The significance of CFD in analyzing the thermal processing techniques is discussed in this chapter. Numerous investigations have been carried out to analyze the velocity profile, flow pattern, temperature profile, location and movement of the SHZ in canned foods. Furthermore, $F_0$ value predictions indicate that the time required for inactivating microorganisms was drastically reduced during the processing of pineapple tidbits compared to pineapple slices. Apart from the canned liquid foods heated primarily by convection, in recent years, research has been extended to canned solid–liquid food mixtures exhibiting combined conduction–convection heat transfer. The effect of can orientation at different inclinations on the temperature distribution inside the can was also analyzed. Geometry modifications have also been included in the survey, exposing the enhanced heat transfer exhibited by upright conical cans. CFD has been valuable in predicting the inactivation kinetics of bacteria and vitamin degradation, which are difficult to tag experimentally during the sterilization process. On the whole, CFD is a precious tool in scrutinizing the various problems associated with the thermal processing industries, and promises to solve all kinds of difficulties that could be encountered in the future.

# Chapter 4
# Computational Fluid Dynamics Modeling for Bread Baking Process

Computational fluid dynamics (CFD) modeling of the entire bread baking process is very complicated, due to the involvement of simultaneous physiochemical and biological transformations. Bread baking is a complex process, where composition, structure and physical properties of bread change along the process, and bread transforms from dough to a product containing soft crumb and crispy crust. CFD can be applied in the modeling of such a complex process. This chapter provides the insight of different radiation models used for modeling of heating in electrical heating ovens; and modeling of the bread baking process, along with the predictions of bread temperature, starch gelatinization and browning index. Moreover, current limitations, recent developments and future applications in CFD modeling of bread baking process are also discussed in detail.

## 4.1 Introduction

Bread is closely related to people's daily life, and bread baking is a food processing technique in which a series of complex physical, chemical and biochemical changes simultaneously take place in the given product. Under the influence of heat, raw dough piece is transformed into a light, porous, readily digestible and flavorful bakery product (Chhanwal et al. 2012). Although bread baking had been in practice for a very long time, comprehensive understanding of the physical process in baking is still lacking due to its complexity (Cauvain 2003). Bread quality predominantly depends on four parameters, i.e. texture, moisture content, bread surface colour and structure (shape and size) of the bread. These four parameters vary during the baking process, due to variations in operating parameters. However, all the above parameters are temperature dependent. Therefore, bread quality can be controlled through optimization of the oven temperature profile (Therdthai and Zhou 2003). Moisture content and temperature are responsible for physiochemical and biological transformations in the bread, such as evaporation of water, gelatinization of starch, volume expansion, crust formation, denaturation of protein and browning reactions etc.; the make the baking of bread a complex process (Therdthai and Zhou 2003). Variations in bread quality can be minimized by proper design of the oven, as well as maintaining proper processing conditions such as air temperature, heating power, baking time and bread size (Zhou and Therdthai 2007; Wong et al. 2007a, b; Chhanwal et al. 2012). Due to the increase in computational power, trained technical people, and availability in the last decade of user friendly and robust software such as Fluent, CFX, and COMSOL Multiphysics etc., use of CFD in the baking industry has increased (Williamson and Wilson 2009).

## 4.2  Bread Baking Process

Baking ovens play an important role in deciding the final product quality of any baking process. Ovens are an integral part of any baking process, and serve as an energy source that leads to heat and mass transfer in the product. The success or failure of any baking venture is determined by the oven and its operating conditions (Chhanwal et al. 2012). Baking ovens can be categorized based upon scale (size and capacity), product being baked (bread or biscuit, etc.), physical arrangements (batch or continuous), heating source (electrical or hot air heating) and mode of heating (conductive or convective). Electrical heating ovens are widely used in the batch baking process, due to their adaptability for different bakery products. The most commonly used continuous bread baking ovens are the travelling tray oven and the rotary rack oven. Location of heating source, air flow, product load, vent position, placement of the bread, and baking time are the major factors affecting heat distribution in the oven chamber. The product quality varies during the initial period of the baking process, due to a high temperature gradient between hot oven and product. Temperature rise occurs due to initial heat absorption, and initiates a number of physical mechanisms (Zhou and Therdthai 2007; Chhanwal et al. 2012).

During the baking process, radiation is the most dominant mode in an electrical heating oven among radiation, conduction and convection heat transfer modes. Radiative heat transfer occurs from red hot heating coils and hot metal surfaces in the form of electromagnetic waves to the surroundings. It occurs by photons, which are emitted by the respective surface and travel in straight lines without attenuation. Heating by radiation depends on emissivity of surface, and the higher the emissivity, the higher the heating rate is (Zhou and Therdthai 2007). Air inside the oven is heated when it comes in contact with the heat source (coils), as well as hot metal parts of oven walls. Heat transfer from hot air to the product surface occurs by convection, and heat from a metal pan is transferred to the bread by conduction. Sparrow and Abraham (2003) extensively studied heat transfer in an electrical heating oven with a variety of geometries, radiative sources and operating conditions of the oven and also placement of thermal load.

Simultaneous heat and mass transfer during bread baking is governed by a mechanism of evaporation–condensation. In this mechanism, water evaporates on the warmer side of the gas cell, absorbing latent heat of vaporization. Water vapor migrates through the gas phase, due to a vapor concentration gradient inside the cell. Further, water vapor condenses, setting free its latent heat on the colder side of the gas cell and becoming liquid. Finally, heat transfer occurs by conduction and moisture is transported by diffusion, through a dough membrane to the warmer region of the next gas cell, where all the above processes reiterate. When the gas phase becomes continuous, the last step no longer exists. This evaporation–condensation mechanism takes place until the temperature gradient remains and the temperature of the whole crumb reaches 100 °C (Therdthdai and Zhou 2003; Wagner et al. 2007; Purlis and Salvadori 2009; Chhanwal et al. 2012).

The baking process occurs as follows: an increase in temperature initiates rapid evaporation of water and release of carbon dioxide, which produces oven spring in the first stage of baking. The top crust is thus pushed up, crumb development follows, and finally, gradual color development occurs. Structural change also occurs during the bread baking process and is comprised of solidification and expansion. Due to the starch gelatinization and protein denaturation, network-like structure forms in the bread's crumb. During gelatinization, starch granules hydrate and swell, and an increase in viscosity, as well as clarity of the medium, is observed. Starch gelatinization is normally used as the minimum baking index. The non-enzymatic browning reaction is induced by increase in temperature and lower moisture content, which results in crust formation at the surface of the bread. Moisture loss due to evaporation of water varies with the properties of dough and baking conditions. Evaporation–condensation of water keeps the moisture of the bread's core region constant, while maximum moisture loss occurs at the surface of the bread. However, the completion of baking results in two different zones within the bread, each with its own unique texture, crumb and crust (Zhou and Therdthai 2007; Chhanwal et al. 2012).

## 4.3 CFD Modeling of the Bread Baking Process

In recent years, CFD has been applied more and more to the design and development of baking ovens, and also to investigate the baking process (Wong et al. 2006; Zhou and Therdthai 2007). The published research works on CFD simulation of bread baking process, along with their salient findings, are summarized in Table 4.1. CFD simulation studies were initiated for the improvement of oven design and optimizing the baking conditions. In an early study, Therdthai et al. (2003) developed a 2D model for an industrial continuous baking oven, to study the temperature and airflow pattern throughout the baking chamber under different operating conditions. Subsequently, the model was extended to a 3D model with moving grid and transient state assumption, to more accurately predict temperature and velocity inside the oven (Therdthai et al. 2004b). The impact of the dough/bread's physical properties on the accuracy of the predictions was analyzed by Wong et al. (2006), using a combined experimental and CFD modeling approach. They found that physical properties have a significant impact, and that density and specific heat capacities tend to be

**Table 4.1** Overview of the research work carried out in CFD simulation of the bread baking process (Chhanwal et al. 2012)

| Description of CFD model | Turbulent/radiation models | Salient findings | Reference |
|---|---|---|---|
| 3D Industrial electrical forced convection oven | RNG $k$-$\varepsilon$ turbulence model | Airflow pattern inside the oven was obtained. Relationship between fan head capacity, fan swirl and oven geometry was studied | Verboven et al. (2000) |
| 2D geometry of industrial continuous bread baking oven. Heat supplied with gas burner and circulated by airflow | Turbulence model with convective heating and steady state simulation | Effect of different energy supply and air volume on oven temperature profile and airflow pattern at various cross sections inside the oven was studied | Therdthai et al. (2003) |
| 3D model with moving grid mesh; U-turn movement simplified by putting into two parts, i.e. front and back part | Turbulence model for transient state simulation | Temperature profile and airflow pattern inside the oven was obtained with tin surface temperature for entire baking process | Therdthai et al. (2004b) |
| 3D geometry of electric heating oven with natural convection flow | S2S radiation model | Temperature profile of baking oven was obtained for broil and baking cycle | Mistry et al. (2006) |
| 2D geometry of industrial continuous bread baking oven was used with sliding mesh. U-turn zone ignored, geometry simplified by flipping lower part and aligning along upper part | DO Radiation with $k$-$\varepsilon$ turbulence model | Effect of thermo physical properties on prediction of bread temperature profile was investigated. Density and specific heat capacities were more dominant factors for accuracy of the model predictions | Wong et al. (2006) |
| 2D geometry with sliding mesh. Grid resolution study was also carried out | DO Radiation model with $k$-$\varepsilon$ turbulence model | Temperature profile of the bread was obtained for entire baking process | Wong et al. (2007a) |
| Incorporation of a UDF for feedback controller to optimize oven temperature | DO Radiation with $k$-$\varepsilon$ turbulence model | Feedback controller tuned fine and preheating time reduced | Wong et al. (2007b) |

(continued)

**Table 4.1**  (continued)

| Description of CFD model | Turbulent/radiation models | Salient findings | Reference |
|---|---|---|---|
| 3D design of gas-fired radiant burner was used in a chamber of industrial tunnel baking oven | Radiation model with shear stress transport $k$-$\omega$ based turbulence model | A burner developed with more uniform radiative flux in oven chamber. Effect of oven chamber humidity and surface emissivity on radiative heat transfer was studied | Williamson and Wilson (2009) |
| 3D geometry of industrial bakery pilot oven was used with controlled air temperature and velocity | S2S radiation model with $k$-$\varepsilon$ realizable turbulence model | Temperature and radiative heat flux were obtained for top and bottom plates | Boulet et al. (2010) |
| 3D domestic electrical heating oven was used with bread. Kinetic model for starch gelatinization was incorporated | Three radiation models, i.e. DO, S2S and DTRM, were used for transient state simulation | The temperature profiles of three radiation models were compared. Starch gelatinization profile for bread slice was visualized | Chhanwal et al. (2010) |
| 3D pilot-scale electrical heating oven containing with nine heating coils was used with bread | DO radiation model | Different temperature profiles of top and bottom chamber were obtained. Effect of placement of bread on baking process time was studied | Anishaparvin et al. (2010) |
| 3D pilot-scale electrical heating oven with bread. Evaporation–condensation mechanism was included for bread baking process | DO radiation model | Crumb temperature remains below 100 °C due to evaporation–condensation mechanism, and thus proper temperature predictions for entire bread obtained | Chhanwal et al. (2011) |
| 3D pilot-scale electrical heating oven with bun | DO radiation model | Velocity and temperature profile differs in partially loaded and fully loaded oven. Quality of bun depends on placement of bun inside the oven | Tank et al. (2012) |
| Heat distribution in an industrial scale oven was studied using CFD modeling | Steady-state Reynolds Averaged Navier–Stokes equations | Improved operating conditions for bread baking with reduced energy usage and baking time | Paton et al. (2012) |

more dominant factors. Wong et al. (2007a) developed a 2D CFD model, extending the earlier works of Therdthai et al. (2003, 2004b) for continuous movement of a traveling tray using sliding mesh technique and employing a discrete ordinate (DO) radiation model. Wong et al. (2007b) proposed the elimination of preheating of the oven before baking, and also incorporated a feedback controller for temperature in the CFD model by using user-defined function (UDF) for the industrial baking oven.

In the literature, different types of radiation models such as discrete transfer radiation model (DTRM), surface to surface (S2S) and discrete ordinate (DO) have been used for analyzing radiative heat transfer inside a baking oven. Mistry et al. (2006) studied CFD modeling of heat transfer in a batch electric heating oven using the S2S radiation model with bake and broil cycle, without the product. Wong et al. (2007a) used the DO radiation model to simulate a continuous baking oven involving u-turn movement with sliding mesh. Dhall et al. (2009) modeled a near-infrared oven using the DO model to predict heating of a cubical food sample. Boulet et al. (2010) used the S2S model to predict radiative heat transfer inside a pilot oven. Recently, Chhanwal et al. (2010) compared all three radiation models (i.e. DO, S2S and DTRM) for a domestic electrical heating oven, and concluded that predictions by all these radiation models were alike. However, the DO model needs more computational power and time compared to the S2S and DTRM. A combination of convective and radiative heat transfer was considered for both glass door and steel walls of the oven, and heating coils were modeled as a volumetric heat source. The authors also studied the temperature profile inside a domestic electrical heating oven and found that a low temperature zone exists near the oven walls.

Anishaparvin et al. (2010) studied the temperature distribution inside a pilot scale electrical heating oven, as shown in Fig. 4.1. A significant difference of temperature profile was observed between the bottom and top tray due to the heating source, inlet vent and product placement. Furthermore, Anishaparvin et al. (2010) developed a CFD model for the process of baking bread in a pilot-scale baking oven, to find out the effect of hot air distribution and placement of bread on temperature and starch gelatinization index of bread. Their study indicated that vent position and placement of bread are the most important factors in influencing the air temperature profile inside the oven cavity. Due to the airflow

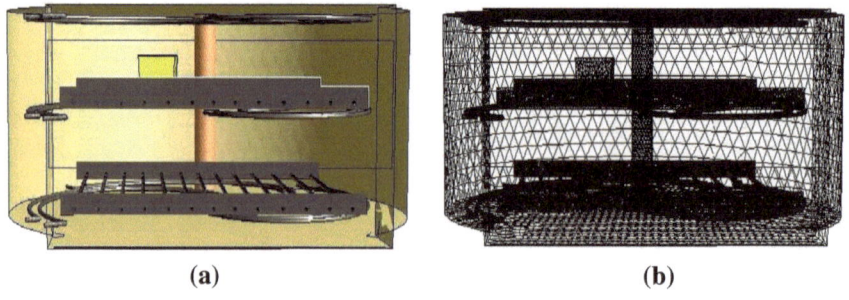

(a)                                                      (b)

**Fig. 4.1**  Oven with bread: (**a**) geometry (**b**) meshed oven (Chhanwal et al. 2011)

pattern, bread placed on the top tray bakes quickly compared to bread placed on the bottom tray. Thus, this study indicates that apart from temperature and time, the placement of bread influences the final bread quality during a batch baking process in an electrical heating oven. Chhanwal et al. (2010) studied the temperature profile of bread during the entire baking process, with preheating of a domestic electrical heating oven.

Chhanwal et al. (2011) used an evaporation–condensation mechanism to imitate the real bread baking process, where bread-crumb temperature remains below 100 °C for the entire baking process. The evaporation–condensation mechanism keeps the bread-crumb temperature in the range of 98–99 °C. However, earlier CFD modeling studies were unable to predict this phenomenon. The evaporation–condensation mechanism was included by defining the specific heat of bread as a function of temperature, including enthalpy jump at the phase change and thermal conductivity of bread as a function of temperature, where thermal conductivity increases until bread reaches 100 °C and is made constant (0.2 W/m–K) once condensation starts (Purlis and Salvadori 2009). In this study, temperature predictions for crumb were identical to the real bread baking process. This model prediction mimics the experimental bread center temperature. This model is very useful for predicting the bread temperature more accurately than the earlier published CFD models. Moreover, this model temperature prediction can also be used to predict physiochemical changes of bread during baking, such as starch gelatinization and browning index.

Recently, Tank et al. (2012) studied the effect of a partially loaded and a fully loaded oven on the temperature profile of a bun, and concluded that placement of the bun inside the oven affects the temperature of the bun and consequently, the quality of the product. Velocity and temperature profiles differ in a partially loaded and a fully loaded oven.

Figure 4.2 shows temperature profiles of a partially loaded oven, in which the top portion of the oven shows a higher temperature as compared to the bottom, due to the natural convection and presence of a vent at the bottom. Higher temperature was noted at the extreme left, close to the oven wall, and in the right central portion, compared to the rest of the oven, which showed non-uniform heat distribution. This non-uniform heat distribution inside the oven was due to the arrangement of the heating source (heating coils), which covered an uneven area within the oven cavity, as well as location of vent. In this study, a 25–30 °C temperature difference was observed in the top and bottom zones of the oven cavity. Due to uneven temperature distribution in the oven, a bun placed on the bottom rack showed slower baking, while a bun placed on the top rack showed rapid baking.

The degree of starch gelatinization can be used as a minimum baking index in the industrial baking process, which decides the baking time of bread. Gelatinization properties depend on the type and origin of the starch. Differential scanning calorimetric (DSC) and X-ray diffraction methods are mainly used for the measurement of the degree of starch gelatinization during the bread baking process. Starch gelatinization is a function of time/temperature and follows first order kinetics (Zanoni et al. 1995a). Zanoni et al. (1995a) studied starch gelatinization using

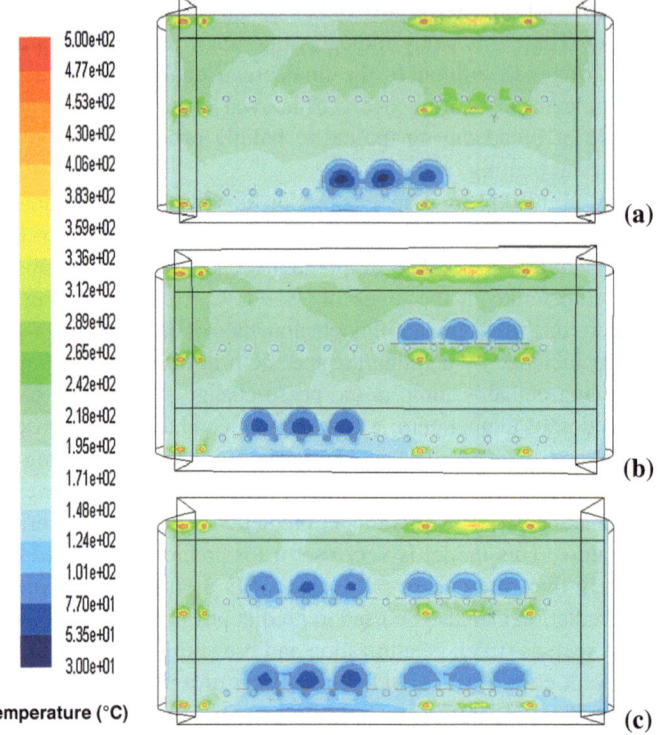

**Fig. 4.2** Comparison of temperature profiles of an oven: (**a**) partially (one tray) loaded (**b**) partially (two trays) loaded, and (**c**) fully loaded oven at 300 s of the baking process (Tank et al. 2012)

the differential scanning calorimetric (DSC) method, and calculated the model parameters such as $k_0$ and $E_a$, as shown in the following equations.

$$I - \alpha = \exp\left(-kt\right) \tag{4.1}$$

where $\alpha$ is degree of starch gelatinization, $k$ the reaction rate constant, and $t$ the time in seconds. The reaction rate constant ($k$) in turn can be calculated using Arrhenius type equation:

$$k = k_0 \exp\left(-\frac{E_a}{RT}\right) \tag{4.2}$$

where $k_0 = 2.8 \times 10^{18}$ (1/s) and $E_a = 138$ kJ/mol (Zanoni et al.1995a).

Change in the temperature profile was correlated with the degree of starch gelatinization in bread, by integrating kinetics model for starch gelatinization with CFD simulation (Therdthai et al. 2004a; Chhanwal et al. 2010; Anishaparvin et al. 2010). Temperature profile and degree of starch gelatinization for a bread slice are depicted in Fig. 4.3.

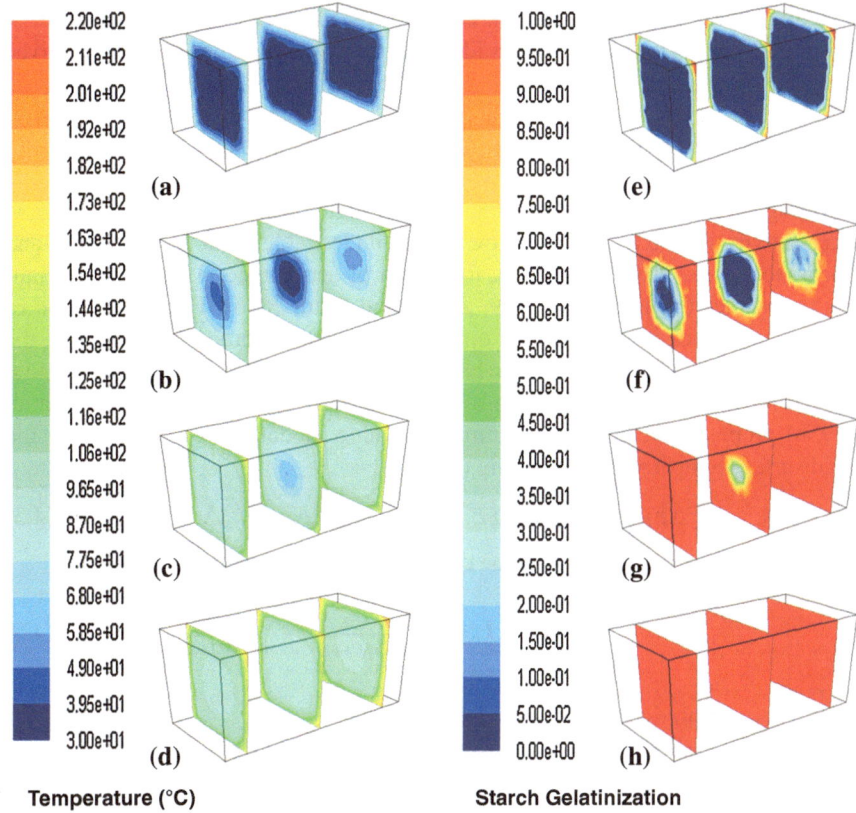

**Temperature (°C)**                                  **Starch Gelatinization**

**Fig. 4.3** Comparison of temperature [(**a**) 60 s (**b**) 300 s (**c**) 600 s, and (**d**) 900 s] and starch gelatinization [(**e**) 60 s (**f**) 300 s (**g**) 600 s, and (**h**) 900 s] profiles in bread during the bread baking process, taken at 7.5 cm to the right and left from center of the bread, in radial direction

Temperature and starch gelatinization profiles of crumb across the bread at (a) 60 s (b) 300 s (c) 600 s, and (d) 900 s at 7.5 cm (left and right) from the center of the bread are depicted in Fig. 4.3a–h. The temperature on and near to the surface of the bread was very high compared to the center, which is responsible for the crust formation and browning of the bread, and thus decides the quality of the bread (Fig. 4.3a–d). Bread core temperature remains below 100 °C, due to the incorporation of evaporation-condensation mechanism. Starch gelatinization progresses from crust to crumb as temperature increases in the crumb region. From Fig. 4.3, it can be observed that the starch gelatinization progressed from crust to the crumb, with the formation of different, corresponding gelatinization zones. The extent of gelatinization at the top and bottom was rapid as compared to the side portion of bread, which is reflected in the temperature profile. Gelatinization was found to be complete within 300 s on outer surface as compared to 900 s at the center of the bread (Fig. 4.3e–h). Starch gelatinization completes within 900 s of baking process, as temperature at center of bread exceeds 80 °C. Formation of different starch gelatinization layers identical to temperature layers were observed inside the bread slice.

Further, computationally simulated temperature can be used to predict the browning of bread surface, as it follows first order kinetics. Browning of the bread surface is, aside from starch gelatinization, an important quality parameter. Browning of the bread surface occurs due to high surface temperature and low water activity (0.4–0.8). Excess browning is undesirable, as it affects flavor and appearance of the bread, as well as formation of acrylamide, which is a carcinogenic substance (Vanina et al. 2009, Zhang and Datta 2006). Thus, evaluation of the browning index is important for optimizing the baking process and, in turn, the quality of bread. Many researchers have proposed various models based on first order kinetics for the determination of browning of bread. However, Zanoni et al. (1995b) proposed the following simple browning model, which depends on the bread surface temperature and baking time:

$$\Delta E_\infty - \Delta E_t = (\Delta E_\infty - \Delta E_{t-\Delta t}) \exp(-k_b \Delta t) \tag{4.3}$$

The reaction rate constant ($k_b$) can be calculated using Arrhenius equation:

$$k_b = k_{b0} \exp\left(-\frac{E_a}{RT}\right) \tag{4.4}$$

where $k_{b0} = 42,000 \ \text{s}^{-1}$, $E_a = 64,151 \ \text{J/mol}$ and $\Delta E_\infty = 52$.

Temperature on the bread surface (i.e. crust), obtained from various points in the bread, are presented in Fig. 4.4. Crust temperature showed a sudden increase (90–110 °C) in the first minute, followed by a linear increase during rest of the baking period. Bread surfaces cross 100 °C within 2 min of the baking process, and this was responsible for the crust formation. Temperatures of bread corners (points 1–7) were higher compared to the surface temperatures (points 8 and 9). Notably, points 3 and 5 showed higher temperatures. Similarly, the temperature on

**Fig. 4.4** Temperature profile of bread crust at different locations during baking process (Chhanwal et al. 2011)

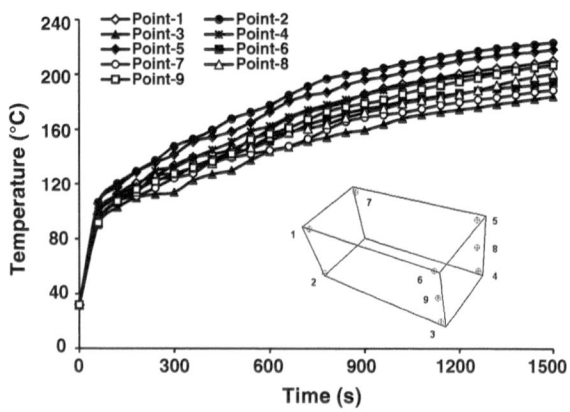

**Fig. 4.5** Browning index
of bread crust at different
locations during the baking
process (Chhanwal et al.
2011)

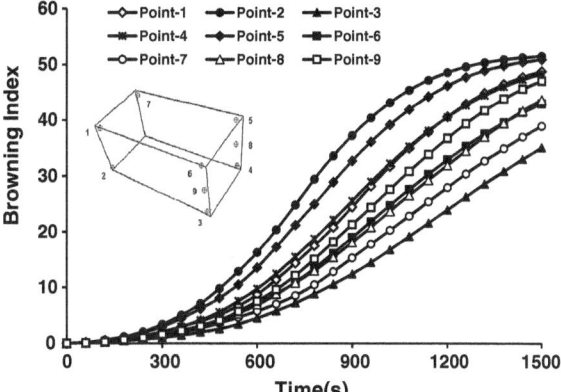

the right side of the bread was higher than on the left side, due to the oven air temperature and airflow pattern inside the oven.

Browning index across bread surface was obtained by integrating a kinetic model with CFD simulation predictions of temperature at various points, and is presented in Fig. 4.5. As the surface temperature of the bread increases, browning also increases. The browning index reached the saturation value (i.e. constant value) toward the end of baking process. The degree of browning at the corner (points 1–7) was rapid compared to the surface of bread (points 8 and 9); this was proportional to the temperatures at those points (see Fig. 4.4).

Zanoni et al. (1995b) determined the maximum browning index (dark brown charred bread) and minimum browning index as 52 and 30, respectively. Figure 4.5 illustrates that browning index of all points fall within this range during the complete baking process. Thus, 25 min of baking was ideal for producing high quality bread under the above baking conditions.

## 4.4  Scope for CFD Modeling in the Bread Baking Process

It is just the beginning of an era of CFD modeling in the process of baking bread, and there is much more to achieve for optimizing different oven conditions and configurations. Although prediction of CFD models match with the experimental observations to some extent, they have some limitations due to assumptions, such as exclusion of moisture transfer and volume expansion. Moisture transfer and volume expansion are critical phenomena in the bread baking process, but due to limitations of commercial CFD software codes, these are not considered in any of the reports published so far. CFD modeling of different heating modes, such as infrared and microwave with or without conventional heating, needs to be further studied for a better understanding of the effects that these heating methods have on the bread baking process.

This chapter highlights the importance of CFD in the modeling of the bread baking process. There is considerable growth in the development and application of CFD simulation in the areas of baking ovens and the bread baking process, to predict the complex flow patterns in the ovens as well as bread temperatures. However, more computational modeling work, including mass transport and volume expansion during bread baking process, needs to be undertaken. Other baking process stages, such as mixing, fermentation, and proofing, need to be thoroughly studied to optimize product quality. On retrospection, this chapter clearly highlights CFD as a valuable tool for the design of baking ovens, and prediction of bread temperature along with various physiological changes such as browning index and starch gelatinization during the bread baking process.

# Chapter 5
# CFD Modeling of Biological Systems with Human Interface

Human digestion is a complex process involving both mechanical breakdown and enzymatic transformation of food. Understanding of this complex digestion process using computational models may enable the food industry to design and provide unique food structure, which helps in efficient conversion and absorption of available nutrients. The stomach plays an important role during the digestion process, and it functions as a reservoir for storing food materials. In addition, it grinds the food material, mixes it with gastric juice and regulates the release of digested materials into small intestine (Kong and Singh 2008). Later, nutrients are absorbed in the human body by enzymatic conversion of partially digested materials (from the stomach) and diffusion through intestinal wall. This chapter discusses the capabilities of computational modeling techniques to provide unique insight on flow behavior inside the human digestive tract (stomach and small intestine). In addition, the effects of structure and material properties of the digesta on delivery of nutrients to the intestinal wall for effective nutrient absorption are also highlighted.

## 5.1 Food Digestion Process

The process of food digestion begins in the mouth, where food is broken down by mastication and lubrication with saliva to form a consistent mass known as a food bolus (Hiiemae and Palmer 1999; Smith 2004). Later on, disintegration of the ingested food bolus takes place inside the stomach, due to the flow of gastric fluid shearing the food surface (Abrahamsson et al. 2005; Kong and Singh 2008). The gastric juice secreted along the walls of the stomach brings about the enzymatic breakdown of food material, while the peristaltic movements generated along the distal part of stomach cause mechanical disintegration. The flow of gastric juice is strongly influenced by peristaltic movements of the stomach wall and rheological properties of the food material. Finally, the discharge of digested materials into the

C. Anandharamakrishnan, *Computational Fluid Dynamics Applications in Food Processing*, SpringerBriefs in Food, Health, and Nutrition, DOI: 10.1007/978-1-4614-7990-1_5, © Chinnaswamy Anandharamakrishnan 2013

small intestine is regulated by the pylorus valve, and the food materials are broken down to particles 1–2 mm in size (Kong and Singh 2010).

## 5.2 Modeling of Food Digestion Inside the Human Stomach

Computational fluid dynamics (CFD) technique can be used to numerically model the dynamics of gastric flow (in terms of pressure and flow fields inside the stomach during peristalsis). It has the capability of mimicking actual mechanical forces acting on food, and thus provides a better understanding of the effect of gastric motility on the extent of mixing of gastric fluid with food during digestion (Pal et al. 2007; Ferrua and Singh 2010; Xue et al. 2012). The following subsection describes stomach modeling processes.

### 5.2.1 Stomach Geometry

The stomach is divided into two compartments: the upper fundus region (reservoir to accommodate incoming ingesta), which occupies one-third of the total stomach, and the remaining lower part, which contains the antrum (grinding and shearing) and pylorus (regulator valve for stomach emptying).

Ferrua and Singh (2010) developed a simplified 3D model capable of reproducing the average shape and capacity of a human stomach (Einhorn 1898; Geliebter et al. 1992; Keet 1993; Kim et al. 2001; Schulze et al. 1998). The model represents the stomach as a "J" shaped organ with a greater curvature of 34 cm, a maximum transverse diameter of 10 cm, a pylorus sphincter diameter of 1.2 cm and a capacity of 0.9 l, as shown in Fig. 5.1.

### 5.2.2 Deformation of Stomach Walls

The movement of the stomach wall known as peristalsis induces mechanical breakdown of food particles. As soon as food material enters the stomach via the esophagus, a series of powerful peristaltic antral contraction waves (ACW) develops within the stomach, propagating toward the pylorus (Ferrua et al. 2011). These antral contraction waves are responsible for the breakdown of food materials by generating fluid motions and mixing gastric juice with food. The patterns for this ACW formation have been described in numerical models developed by Pal et al. (2004) and Ferrua and Singh (2010). ACW initiated every 20 s at a distance of 15 cm from the pylorus, and their velocity of propagation is about 2.5 mm/s (Fig. 5.2).

The total lifespan of an ACW is 60 s. The motility pattern is periodic after 40 s and repeats every 20 s (Xue et al. 2012). The depth of this contraction has been

**Fig. 5.1** Three dimensional
geometry of the human
stomach (Ferrua et al. 2011)

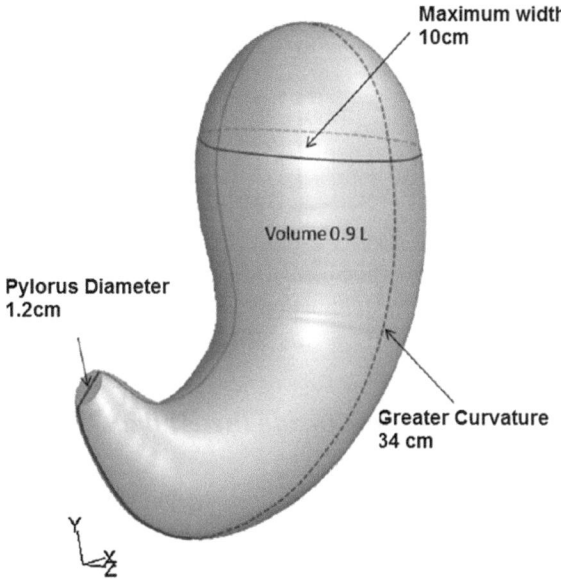

**Fig. 5.2** ACW motility
pattern in the stomach (Xue
et al. 2012)

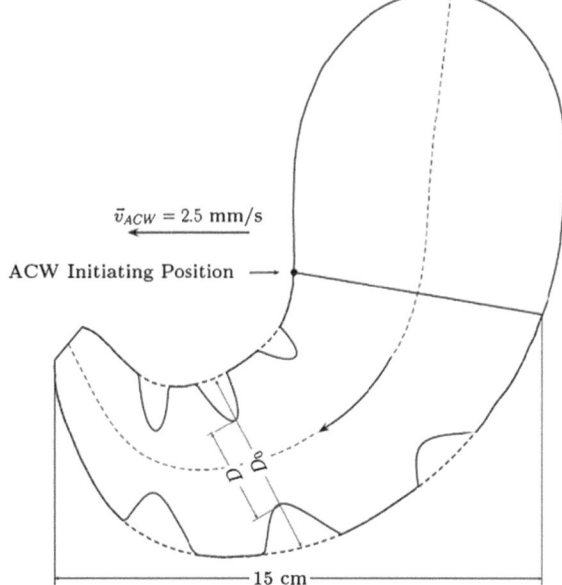

defined based on an occlusion ratio ($\varepsilon$), which is the ratio of change in diameter of
section before and after contraction (Pal et al. 2004; Ferrua and Singh 2010). The
length of propagation, amplitude of contractions, and the time taken by the stomach to generate the contraction waves are affected by the stomach volume, as well

as by chemical and physical properties of the meal (Camilleri and Prather 1993; Mayer 1994). A typical ACW pattern is defined in Eq. 5.1 (Pal et al. 2004):

$$\varepsilon = 1 - \frac{D}{D_0} = \begin{cases} 0.25 - 0.25(17.5 - t)/17.5 & 0 \le t \le 17.5 \\ 0.25 & 17.5 \le t \le 33.5 \\ 0.5 - 0.25(57 - t)/23.5 & 33.5 \le t \le 57 \\ 0.5 - 0.5(t - 57)/3 & 57 \le t \le 60 \end{cases} \quad (5.1)$$

where '$D$' is the diameter of the contraction ring, '$D_0$' is the original diameter of the section before contraction, and $t$ is the time in seconds.

## 5.2.3  Fluid Flow Inside the Human Stomach

The flow of gastric juice inside the human stomach is predominantly character-ized by two flow patterns; namely, retropulsive jet-like motion and flow recir-culation, i.e. formation of eddies (Ferrua et al. 2011). These two flow patterns are mainly responsible for the mixing and grinding of food particles inside the human stomach (Pallotta et al. 1998; Pal et al. 2004; Schulze et al. 1998). The contraction force of the wall directly elevates the pressure of the stomach in a closed system. The pressure difference between the two compartments—the body and the antro-pyloric region inside the stomach—creates the vortex flow. As the contraction waves propagate toward the pylorus, the pressure builds up in the distal part of stomach, i.e. the antro-pyloric region. This increase in pressure at the pylorus region causes the food material to be forced back into the mix-ing region (i.e. the antrum). This particular flow of gastric juice that causes food material to be forced back to the antrum region is called retropulsion (Ferrua and Singh 2010).

## 5.2.4  Numerical Equations Governing Fluid Flow

There are two different modeling approaches used to model the multiphase flow (involving more than two phases) dynamics occurring in the stomach. Firstly, the Euler–Lagrange approach is used to track the movement of each particle by taking particle–particle interactions and the forces acting on them into account. Secondly, the Euler–Euler approach considers all phases to be continuous and interpenetrating. As the former method is expensive and not suitable for dense particulate suspension, the Eulerian model was found suitable for gastric fluid modeling (Xue et al. 2012). The flow field that develops within the stomach was modeled with a laminar and incompressible fluid flow of a continuous liquid phase (Pal et al. 2004, 2007). Under these flow conditions, the conservation of mass and momentum within the system is given by Eqs. 5.2 and 5.3, respectively (Ferrua and Singh 2010).

$$\frac{\partial u}{\partial x} = 0 \tag{5.2}$$

$$\frac{\partial u}{\partial t} + u \cdot \nabla u = -\frac{1}{\rho}\nabla p + \frac{\mu}{\rho}\nabla^2 u \tag{5.3}$$

where '$u$' is the velocity component (m/s), '$p$' is the pressure (Pa), '$\mu/\rho$' is the kinematic viscosity of the fluid (cP), and '$\rho$' is the density of the fluid (kg/m$^3$).

## 5.3  Rheological Properties of Food Materials

The rheological properties of the fluid can have a significant effect on the dynamics of gastric flow during digestion. Efficient digestion of ingested materials can be achieved through appropriate alterations in the physical structure and reduction in the content of non-nutrient components that influence the viscosity of the liquid phase (Ferrua et al. 2011).

### 5.3.1  Effect of Viscosity on Characteristic Flow Field Within the Stomach

Ferrua and Singh (2010) studied the effects of viscosity changes on the flow behavior of gastric juice based on the two characteristic flow patterns discussed earlier. Figure 5.3a–b shows the prediction of a characteristic flow field for a Newtonian fluid at two different viscosities.

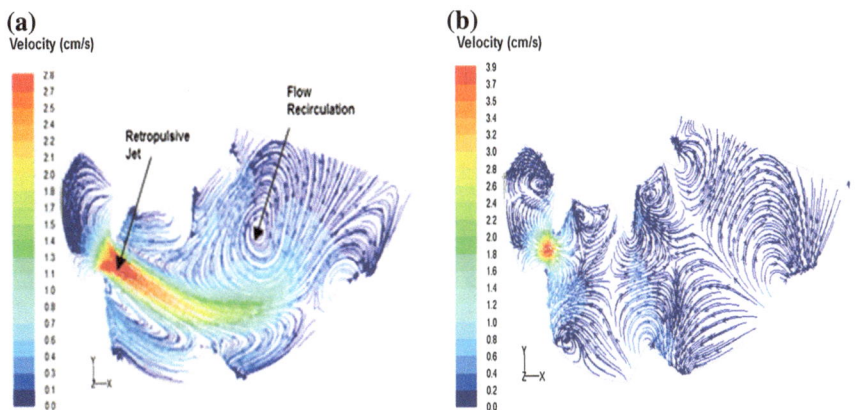

**Fig. 5.3** Gastric fluid velocity profile (cm/s) associated with (**a**) 1 cP viscous fluid and (**b**) 1,000 cP viscous fluid (Ferrua and Singh 2010)

It can be observed from Fig. 5.3 that gastric fluid with 1cP viscosity attained a maximum flow velocity of 2.8 cm/s (Fig. 5.3a), and that when viscosity increased to 1,000 cP, the maximum velocity attained was 4 cm/s (Fig. 5.3b). The results were validated with previously published experimental results, and suggested that an increase in fluid viscosity diminished the formation of a retropulsive jet (as seen in Fig. 5.3b) and resulted in poor mixing of food particles with gastric juice (Ferrua and Singh 2010).

## 5.4 Effect of Solid–Liquid Density Difference on Particle Distribution

The density difference between two phases may impact particle distribution within the mixture through buoyancy effects (Xue et al. 2012).

From Fig. 5.4a–b, it can be seen that when the particle density was increased to 5 % greater than the gastric fluid, a retropulsive jet with a very high maximum velocity of 6.2 cm/s (Fig. 5.4b) was observed near the antro-pyloric region. This may be due to sedimentation of heavier particles at the bottom of the stomach, resulting in narrower passage for fluids to flow through. This leads to the creation of a higher retropulsive velocity jet when compared to single phase.

## 5.5 Effect of Particle Loading on Mixing

Xue et al. (2012) studied the effect of particle loading ratio on the mixing of food particles using the multiphase Euler–Euler modeling approach. The particle volume loading that defines the average distance between individual particles has been found at various levels, as shown in Fig. 5.5a–d.

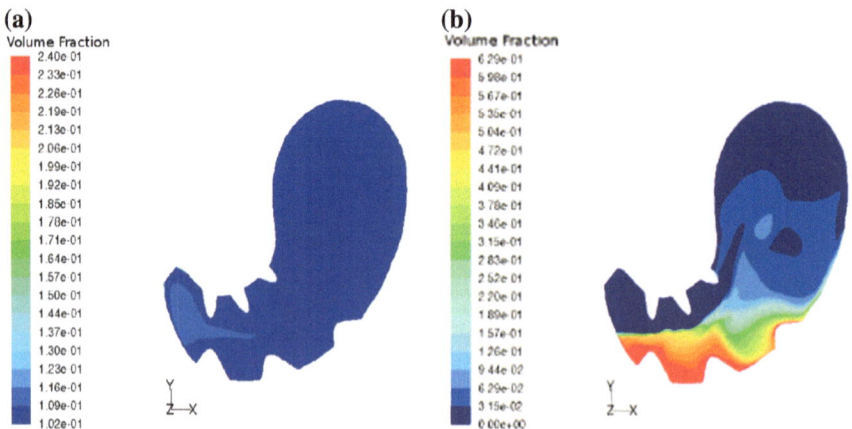

**Fig. 5.4** Particle volume fraction (%) contour in the stomach with mild ACWs at time $t = t_0 +10$ s with (**a**) $\rho = 1,000$ kg/m$^3$ and (**b**) $\rho = 1,050$ kg/m$^3$ (Xue et al. 2012)

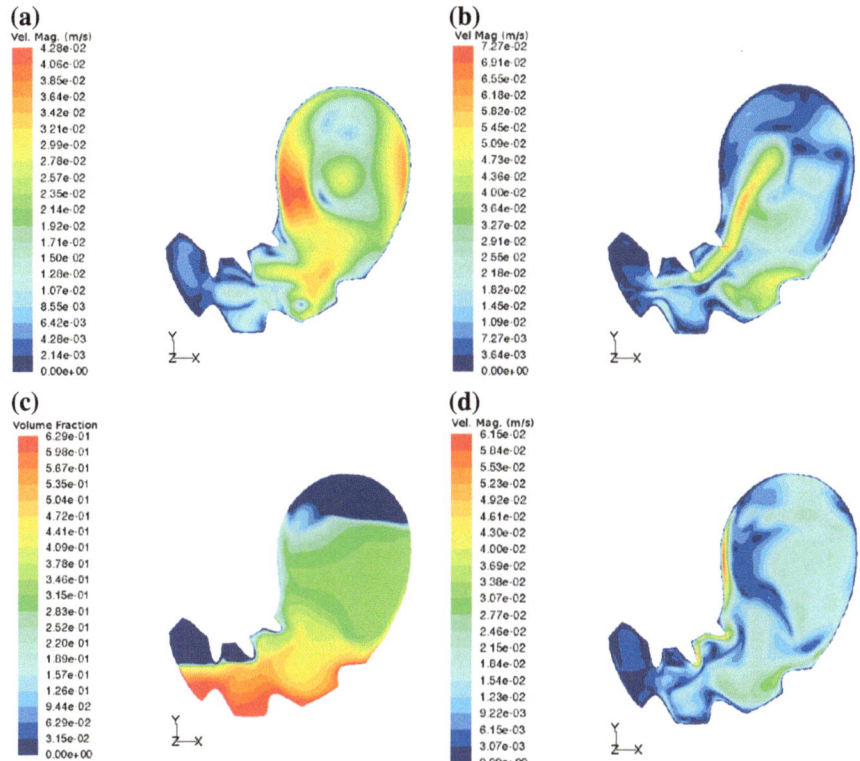

**Fig. 5.5** Velocity magnitude for gastric juice (m/s) with 2 mm particle diameter and 10 % particle volume fraction at (**a**) particle density, $\rho = 1{,}000$ kg/m$^3$, and (**b**) $\rho = 1{,}050$ kg/m$^3$; when particle volume fraction increased to 30 % with 1 mm diameter at (**c**) $\rho = 1{,}050$ kg/m$^3$ and (**d**) $\rho = 1{,}050$ kg/m$^3$, respectively (Xue et al. 2012)

The velocity profile for gastric juice at 10 % (Fig. 5.5a–b) and 30 % (Fig. 5.5c–d) particle loading levels were analyzed after 10 s of processing time. The authors found that there was a significant reduction in mixing capability of ACW at a higher particle loading ratio of 5 % difference between solid and liquid phases. From Fig. 5.5, it can be observed that a higher particle volume fraction can block the passage to form a retropulsive jet, resulting in a reduced degree of mixing.

## 5.6  Modeling of the Absorption Process in the Small Intestine

The small intestine is a complex and highly regulated system. It is the longest part of digestive tract system, measuring about 2–6 m in length. The total surface area of the human small intestine is $2 \times 10^6$ cm$^2$ (Mahler et al. 2012), and the duodenum is the first section of the small intestine, followed by the jejunum and ileum (Fig. 5.6), comprising 5, 50, and 45 % of the small intestinal length, respectively

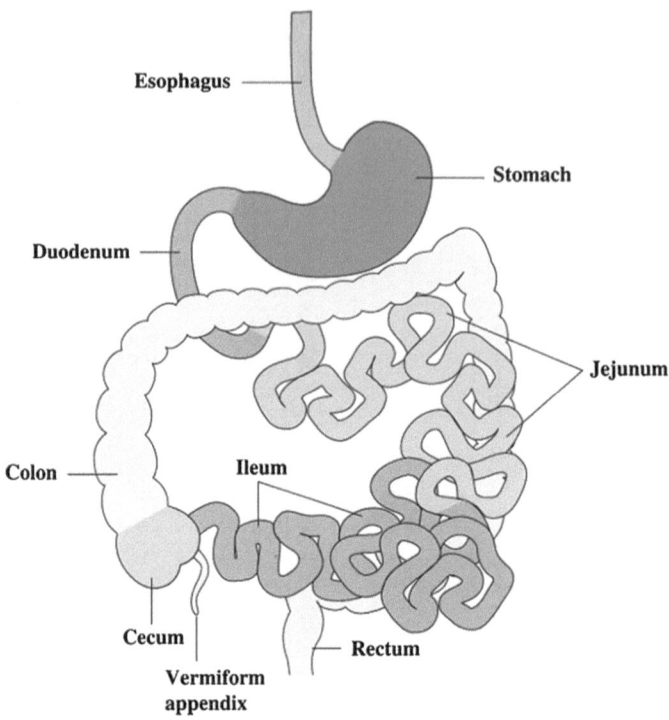

**Fig. 5.6**  Three parts of the small intestine (Desesso and Jacobson 2001)

(Balimane and Chong 2005). The diameter of gut varies along each section of the small intestine and is considered to be an average of 2.5 cm, varying from 5 cm in the upper duodenum to less than 2 cm in the distal ileum (Stoll et al. 2000).

Modeling of the small intestine is to obtain efficient conversion and increased bioavailability of nutrients; there should be perfect mixing of gastric contents inside intestinal lumens to enable faster diffusion of nutrients through the wall. Development of such a model has been a great challenge due to the complexity involved in mass balance around the small intestine (Stoll et al. 2000; Spratt et al. 2005; Dokoumetzidis et al. 2007). It is very complex to understand the mechanisms of digestion and absorption inside the small intestine, since there is very little information available from an engineering perspective (Norton et al. 2006; Tharakan et al. 2007). Computational modeling can be used for understanding the absorption pattern in the small intestine with the effects of physiological variables, physicochemical properties and their formulation (Parrott and Lave 2010). Using the Computational Fluid Dynamics (CFD) and/or Finite Element Modeling (FEM) technique, it is possible to theoretically evaluate fluid dynamic parameters such as intestinal fluid velocities and shear rates during the segmentation motion of the small intestine.

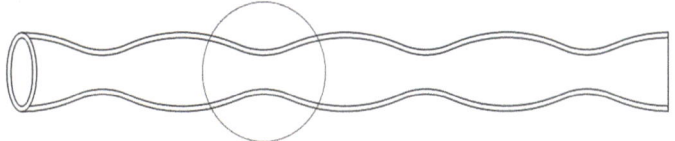

**Fig. 5.7** Schematic representation of the segmentation motion of the small intestine

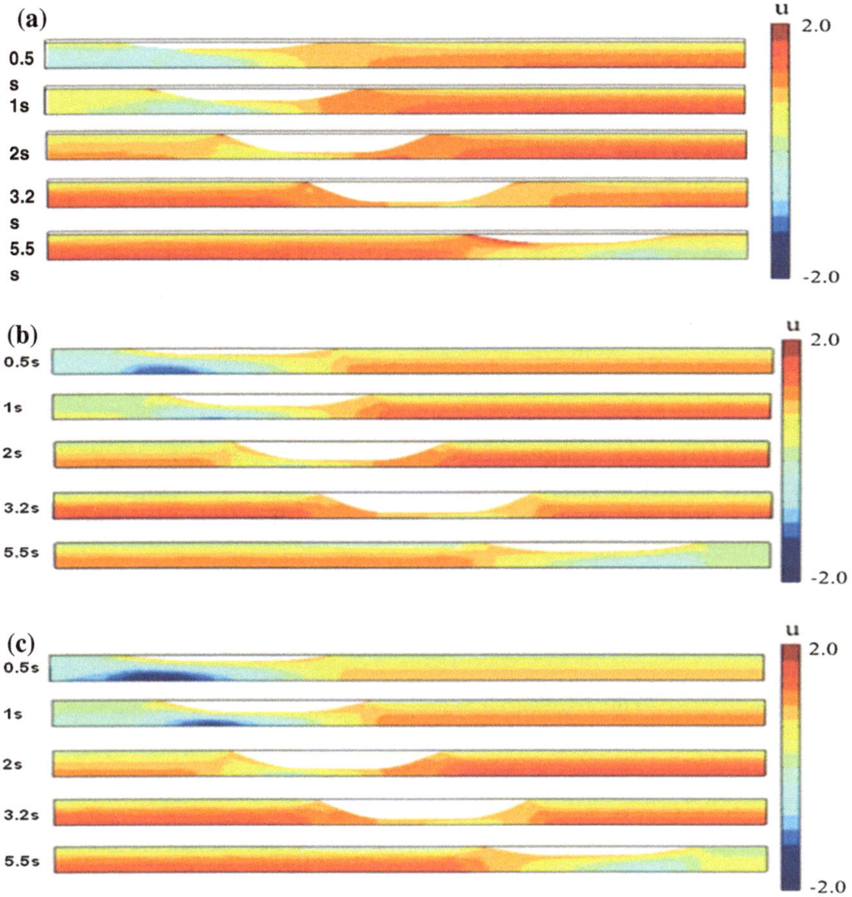

**Fig. 5.8** Horizontal velocity profiles (cm/s) for flow of Newtonian fluids with (**a**) $10^{-3}$ m$^2$ s$^{-1}$ (**b**) $10^{-4}$ m$^2$ s$^{-1}$, and (**c**) $10^{-5}$ m$^2$ s$^{-1}$ viscosity (Love et al. 2012)

## 5.6.1 Movements in the Small Intestine Causing Mixing of Food

The movement of the small intestinal wall is responsible for the flow behavior of intestinal fluids, and can be divided into segmentation contractions and propulsive contractions (Guyton and Hall 2006). The propulsive contractions, known as peristalsis, occur on successive sections of circular smooth muscles and are responsible for propelling the digesta through the small intestine. This involves approximately 3–6 h of transit time (Liu et al. 2003), with an average velocity of 1.0 cm per minute inside the intestine (Guyton and Hall 2006).

Segmentation contractions are primarily responsible for the mixing of digesta with surrounding fluid. This begins as soon as the intestine is filled with digesta, after which localized concentric contraction takes place in the gut, as shown in Fig. 5.7.

## 5.6.2  Effect of Wall Contractions on Flow of Intestinal Contents

Love et al. (2012) used a finite element model to simulate flow through the small intestine as induced by peristalsis and segmentation motion. The segment of intestine was represented by a two dimensional mesh, and equations were formulated to describe the motion of the mesh boundary during peristalsis.

Figure 5.8a–c shows the horizontal velocity profiles for flow of fluids with various viscosities through a tube simulating the small intestine. The model was solved using a finite element method to determine the consequent deformation of the mesh, as a function of time. It was assumed that the flow of digesta could be adequately represented by a 2D cross-section of intestine with a length of 10 cm and a diameter of 0.5 cm.

CFD modeling of biological systems allows the food industry to design and provide unique food structure, as both mixing and food formulation have significant effects on mass transfer phenomena. At the same time, as mixing enhances mass transfer processes, increase in viscosity of the food results in a reduction of the absorption across the membrane of small intestine.

# Chapter 6
# Computational Fluid Dynamics Modeling for High Pressure Processing

High pressures processing (HPP) has shown its importance in food preservation, due to its microbial inactivation ability at lower temperature, i.e. non-thermal sterilization, and high pressure in the range of 100–900 MPa. HPP keeps a product's freshness intact, along with nutrients such as vitamins. HPP is used for pressure shift freezing, high pressure thawing, and high pressure thermal (HPT) processing. HPP technology has shown rapid development in the preservation and processing of liquid and high moisture foods, such as fruit juices, ready-to eat meals, meat products, sea-food products, etc. Combined with high temperature, high pressures thermal (HPT) processing has shown promise in increasing shelf-life of foods at room temperature for months to years. HPT helps to achieve commercial sterile low acid shelf-stable food products. However, the temperature profile of a food during the high pressure thermal process has a significant effect on the final product's quality and preservation. Therefore, it is very important from the food safety point of view to achieve uniform temperature inside the processing vessel (Koutchma 2012; Ghani and Farid 2007a).

Product composition plays an important role in optimizing processing conditions for the HPT process, as temperature increase per 100 MPa varies with product composition. Earlier, researchers observed that during the HPP process, the rate of microbial inactivation decreased with processing time. Therefore, for product safety and quality, microbial inactivation should be ensured by considering complicated intrinsic and extrinsic factors. Computational Fluid Dynamics (CFD) modeling can help in optimization of processing conditions and system structure of the HPP process (Koutchma 2012; Ghani and Farid 2007a, b). In the last decade, CFD modeling has shown promise in predicting the temperature of canned product during thermal sterilization and pasteurization processes. Similarly, CFD can be used to predict the temperature profile across a vessel during HPT processing.

Generally, high pressure processing is carried out in batch mode, which consists of compression phase, pressure-holding phase, and pressure release phase. In compression phase, pressure is increased to the desired level. This is followed by

C. Anandharamakrishnan, *Computational Fluid Dynamics Applications in Food Processing*, SpringerBriefs in Food, Health, and Nutrition, DOI: 10.1007/978-1-4614-7990-1_6, © Chinnaswamy Anandharamakrishnan 2013

the pressure-holding phase, where the desired elevated pressure is kept constant for a few minutes. Lastly, the pressure is reduced to ambient pressure. In the HPP system, water is usually used as a pressure-transmitting medium, i.e. carrier material. During HPT, heat is generated volumetrically within the food due to the rapid pressurization. HPP has several advantages over conventional processes, such as retention of freshness, texture, flavor, and color, along with reduced processing time. Commercially available HPP products include oysters, fruit preserves, meats, and milk (Ghani and Farid 2007a; Khurana and Karwe 2009).

Earlier, Knoerzer et al. (2007) developed a model for a pilot scale 35 L vessel for the HPT process using polytetrafluoroethylene (PTFE) carrier. PTFE carrier helps to produce thermal uniformity in a vessel, and is thus able to produce 12D reduction during the process, whereas an empty vessel fails to achieve spore reduction. Further, Juliano et al. (2009) used a computational thermal fluid dynamic (CTFD) model to predict sterilization uniformity in a 35 L HPT vessel by predicting inactivation of *C. botulinum*. A CFD model for high pressure compression of beef fat and water mixture was developed to obtain the temperature distribution along pressure and velocity profiles. It was observed that solid pieces of food were heated more rapidly than liquid food, due to a different compression-heating coefficient (Ghani and Farid 2007b).

Khurana and Karwe (2009) developed a transient model of high hydrostatic pressure processing to predict the temperature distribution of a pressurizing medium (water). Water was pressurized in a 10 L vessel of a high hydrostatic

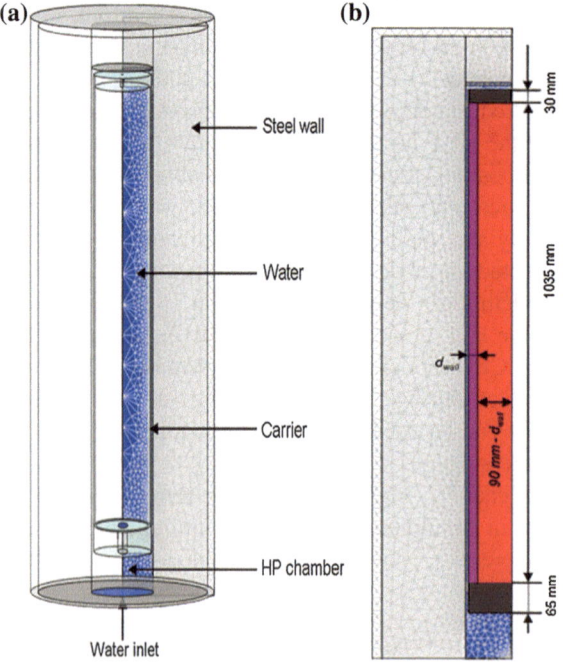

**Fig. 6.1** Configuration of model geometry and mesh in a vertical cross section of the high-pressure vessel steel structure containing the PTFE carrier. (a) HP system (b) HP system with carrier dimensions indicated (Knoerzer et al. 2010)

pressure processing unit, from 0.1 to 586 MPa in 180 s, with an initial temperature of 25 °C. Pressure was maintained for 600 s in the pressure hold up period.
It was observed that adiabatic heat generation in water and cooling at the walls
result in non-uniform temperature distribution at the end of the process. It was
also observed that the increasing initial temperature of the pressurizing medium
showed an increase in temperature non-uniformity. Vessel wall temperature distribution was affected, mainly due to natural convection and conjugate conduction
heat transfer. However, due to the unavailability of water properties (such as density,
heat capacity, thermal conductivity, viscosity, and thermal expansion coefficient)
at elevated pressure, all these properties were assumed to be constant. Due to this
assumption, the model predictions were not very accurate.

Knoerzer et al. (2010) developed a model for carrier optimization in a pilot-
scale high-pressure sterilization plant using finite element methods. The geometry
of the model of the high-pressure chamber is shown in Fig. 6.1a, and meshed
geometry is depicted in Fig. 6.1b. In Fig. 6.1a, the outer cylinder shows the high-
pressure chamber, the inner cylinder shows the carrier, and all shaded areas represent
the axis-symmetrical computational domains (Knoerzer et al. 2010). In this model,
the k-ε turbulence model was used for the water domain, with the convection and
conduction model applied to both liquid and solid regions. Variations in the physical properties of water (expansion coefficient, density, specific heat capacity, thermal
conductivity and viscosity) as a function of temperature and pressure were obtained
from NIST/ASME database. It was found that the physical properties follow a first
order polynomial equation, with temperature and pressure as variables (Knoerzer
et al. 2010).

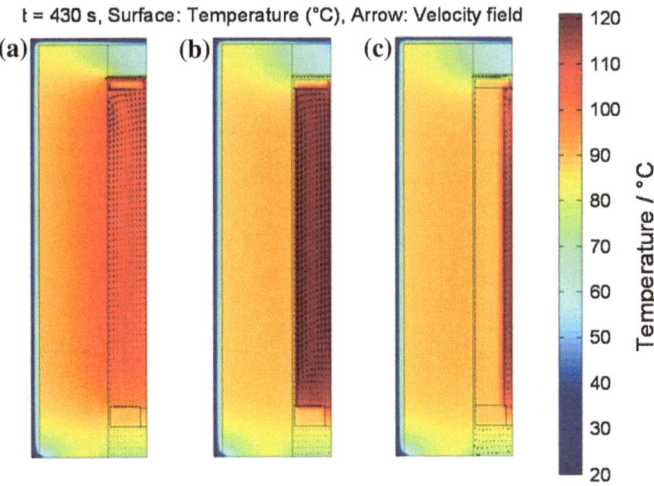

**Fig. 6.2** Examples of simulated temperature distributions in the HP system for three different
wall thicknesses (**a**) 0 mm (**b**) 5 mm (**c**) 70 mm at the end of hold time ($t = 430$ s) (Knoerzer
et al. 2010)

Figure 6.2 depicts temperature distribution for three different wall thicknesses (0, 5, and 70 mm) at the end of pressure hold time (430 s). In the case of no carrier wall thickness, temperature does not cross 105 °C, due to lack of insulation (Fig. 6.2a). For the 5 mm wall thickness, the average temperature reached 118 °C, with uniform temperature distribution at the end of pressure holding time (Fig. 6.2b). Therefore, insulation thickness has a major role in HP processing (Knoerzer et al. 2010). For the 70 mm wall thickness, the effective volume of vessel reduced to 1.3 L. Carrier walls insulate the contents from the cooler vessel and act as heat sinks. This resulted in a temperature gradient and also heat loss (Fig. 6.2c). With the help of CFD modeling, in the near future it will be possible to optimize different carrier materials and their thicknesses, which will enhance performance of the HPT processing system. Recently, Knoerzer and Chapman (2011) highlighted the importance of accurate input data, such as thermo–physical properties of a material and their correlation to pressure and temperature and process conditions, for precise model predictions for the HPT process.

All the above studies clearly indicate that CFD can be effectively used for the modeling of high-pressure processing. CFD modeling helps in designing and optimizing the carrier wall thickness for optimum uniform temperature distribution and also maximum usable volume. However, lack of thermo-physical properties data for different material and processing conditions is a major hurdle in the precise modeling of high-pressure processing.

# Chapter 7
# Applications of Computational Fluid Dynamics in Other Food Processing Operations

CFD simulations can be used as a tool to improve the spray-freezing operation by determining particle velocities, temperatures, and impact positions on the wall during the spray-freezing process for different equipment configurations and operating conditions. Jet impingement heating is more superior to traditional convection heating due to its uniform heating and high heat transfer rate with less baking time. CFD modeling can be used for design and development of impingement jet oven. CFD modeling has a huge potential to address the problems facedby large scale grain processing industries. It is possible to model unit operations in the flour milling industry, where problems in handling and operating the equipment are encountered. This chapter discuss the applications of CFD in spray-freezing operations, jet impingement ovens, spouted bed and flour milling industry.

## 7.1 CFD Simulation of Spray Freezing Operations

Freeze-drying is a popular method of producing shelf stable particulate products, and is of particular value for drying thermally sensitive materials (usually biologically based), which can be heat damaged by higher temperature methods, such as spray drying. Porous structures are formed from the creation of ice crystals during the freezing stage, which subsequently sublime during the drying stage, and this often leads to good rehydration behavior of the product. It is possible to produce freeze-dried product in powdered form using a technique known as spray freeze-drying (Malecki et al. 1970; Heldman 1974), in which a liquid stream containing a dissolved solid is atomized in a manner similar to spray drying, then brought into contact with a cold fluid to freeze the droplets. These are finally freeze-dried, either conventionally or in a fluidized bed (Leuenberger et al. 2006; Claussen et al. 2007; Anandharamakrishnan et al. 2010b). One method of spray freezing is by contacting with a cold gas. This is a complex process that involves a number of mechanisms: (1) the formation and the motion of individual drops with respect to

D. Anandharamakrishnan, *Computational Fluid Dynamics Applications in Food Processing*, SpringerBriefs in Food, Health, and Nutrition, DOI: 10.1007/978-1-4614-7990-1_7, © Chinnaswamy Anandharamakrishnan 2013

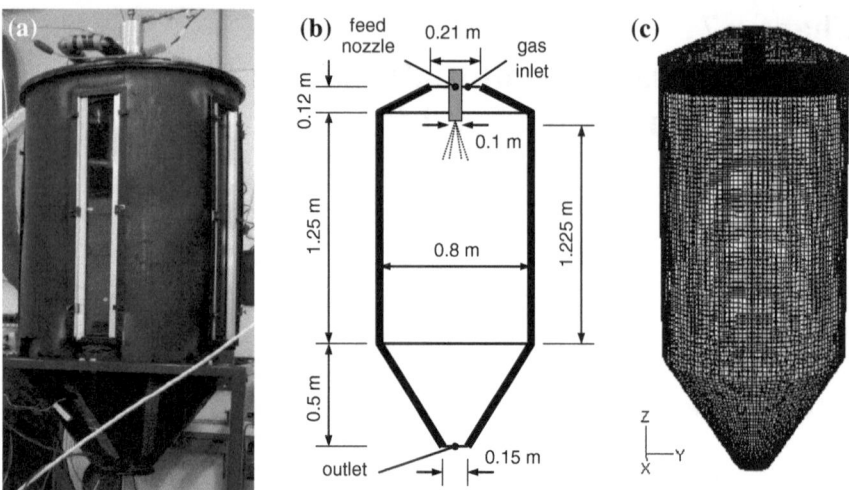

**Fig. 7.1**  Spray freezing chamber (**a**) photograph and (**b**) dimensions; (**c**) computational domain and meshing (Anandharamakrishnan et al. 2010b)

each other and the gas is determined by the fluid mechanics of the spray; (2) the heat transfer between the gas and the droplets depends on local conditions, e.g. gas temperature, droplet temperature and droplet-gas slip velocity; and (3) the freezing and ice crystallization within the drops.

### 7.1.1  CFD Simulation Methodology

Anandharamakrishnan et al. (2010b) used the geometry shown in Fig. 7.1a–c, with a solid cone spray pressure nozzle atomizer located near the top of the chamber; the freezing gas (liquid nitrogen) enters via an annulus. The standard k-ε turbulence model was used, with inlet k and ε values calculated according to Langrish and Zbicinski (1994). The "escape" wall boundary condition (where particles are lost from the calculation at the point of impact with the wall) was used. In the 3D model, a hexahedral mesh was used (typical size is 0.001 m) with 180 k grid cells (preliminary tests with a finer grid showed that 180 k cells were sufficient to obtain grid independent solutions for the mean velocity field). The grid geometry is shown in Fig. 7.1c. To maintain the accuracy of the solution near the nozzle, a fine mesh was used.

The CFD code Fluent 6.3 was used to simulate in 3D the co-current flow spray freezing unit fitted with a pressure nozzle with a solid cone spray for Case A and Case C, and a hollow cone spray for Case B. The finite volume method was used to solve the partial differential equations of the model using the Semi-Implicit Pressure-Linked Equations (SIMPLE) method for pressure–velocity coupling and a second-order upwind scheme to interpolate the variables on the surface of the control volume. Two-way coupling between the cooling medium and 'inert particles' using the discrete phase model (DPM) was used; the stochastic effects of the turbulence on the particle trajectories were included through an eddy-interaction model (see details in Anandharamakrishnan et al. 2010b).

The heat transfer between the droplet and the cold gas was computed based on the following equation:

$$m_p c_p \frac{dT_p}{dt} = hA_p \left(T_g - T_p\right) \tag{7.1}$$

where $m_p$ is the mass of the droplet, $c_p$ is the droplet specific heat (J/kg K), $T_p$ is the droplet temperature (K), $A_p$ is the surface area of the droplet (m$^2$), and $h$ is the heat transfer coefficient (W/m$^2$K) (Fluent, 2006).

The current Fluent 6.3 DPM model does not include phase change during freezing (solidification). Single droplet freezing studies (Hindmarsh et al. 2003; MacLeod et al. 2006) indicate that freezing comprises a number of stages: (1) super cooling to below the normal freezing temperature; (2) nucleation; (3) recalescence, whereby rapid crystal growth occurs with a sudden temperature rise, as crystal growth liberates latent heat and the droplet warms up to the normal freezing temperature; (4) further, slower crystal growth that is limited by heat transfer from the gas, during which some freezing point depression may occur; and (5) once freezing is complete, cooling of the frozen particle to the gas temperature. The effect of the latent heat of fusion during the recalescence and subsequent growth stages (3–4) was approximated in the model by assuming that solidification takes place linearly over a temperature range between 0 and −10 °C (Anandharamakrishnan et al. 2010b).

## 7.1.2 Comparison Between Measured and Predicted Gas Temperatures

The experimental temperature profiles of the gas are plotted in Fig. 7.2a–c at axial positions of $z = 0.58$, 0.8 and 1.23 m below the nozzle. When the CFD model of cooling gas is compared to the experiment, the cooling gas flow pattern appears to show some difficulty in penetration through the spray region. In the experiment, however, there may be droplets of liquid nitrogen entrained in the inlet gas flow, which is able to penetrate into the core of the spray and provide significant cooling, the effect of which is not included in the CFD model. Further down the chamber at $z = 0.8$ and 1.23 m (where the conical section begins), the temperature profiles flatten and the core region of higher temperature broadens as the spray fans out due to (1) evaporation of liquid nitrogen and (2) super cooling/recalescence. The temperature of gas set outside the core is almost uniform, and it appears that most of the droplets do not penetrate into this zone. This trend was also observed by Kieviet (1997) and Huang et al. (2006) in their spray drying studies.

**Fig. 7.2** Comparison of gas
temperature profiles between
measurements and CFD
predictions for Case A (solid
cone spray) at (**a**) $z = 0.58$ m
(**b**) 0.8 m, and (**c**) 1.23 m
from the nozzle spray point
(Anandharamakrishnan et al.
2010b)

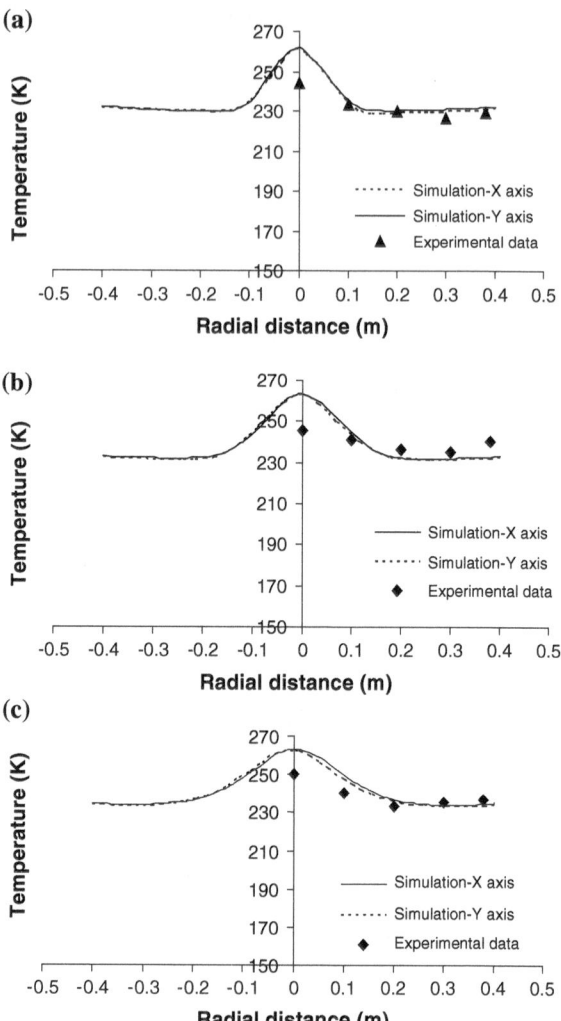

Particle trajectories are shown in Fig. 7.3, and clearly reveal that some particles
are re-circulated by the gas phase and have upward velocities close to the walls.

## 7.1.3 Particle Impact Positions

Knowledge about the positions where particles are having impact is important
for designing and operating spray freezing equipment. Comparisons of the simulated and experimental results for particle impacts on the chamber walls are shown
in Fig. 7.4a–b (top views) and c–d (front views). These figures indicate that a
large fraction of the particles (65 %) strike the conical part of the spray freezing

**Fig. 7.3** CFD simulated
particle trajectories
(Anandharamakrishnan et al.
2010b)

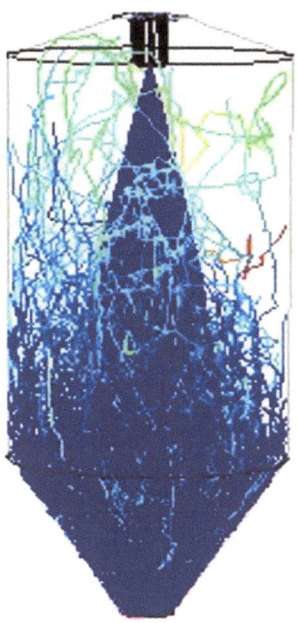

chamber; 11 % of the particles hit the cylindrical part of the wall, and only a small proportion (22 %) of the particles come directly out of the chamber. The interesting observation is that no particles hit the ceiling due to gas re-circulation, which happens only at the bottom (see Fig. 7.3). These results were in reasonably good agreement with the experimental observations shown in the photographs of Fig. 7.4b–d (Anandharamakrishnan et al. 2010b). An important point to note is that in these experiments, a significant number of particles stick to the walls.

In spray drying simulations, it is generally assumed that (non-sticky) particles slide down the walls toward the main product outlet. However, in spray freezing, a little more attention is required, as when a frozen particle hit the wall during the operation, it tended to stick and build an icy layer, as shown in Fig. 7.4b. This may be because either the ice particles (being crystalline) are rougher, or because incomplete freezing has occurred (which is likely, considering the results for large particles). Hence, in order to maximize the freezing efficiency, the amount of product conveyed to the outlet, as well as maintaining a sufficiently cold wall temperature during the process, needs a little more consideration.

## 7.2  CFD Modeling for Jet Impingement Oven

Success and growth of any industry depends on good quality product and energy efficiency of the process. In the food industry, heating is one of the main process techniques used for enhancement of product shelf-life, flavor, texture, palatability etc. Conventional heating or cooling methods give good quality product.

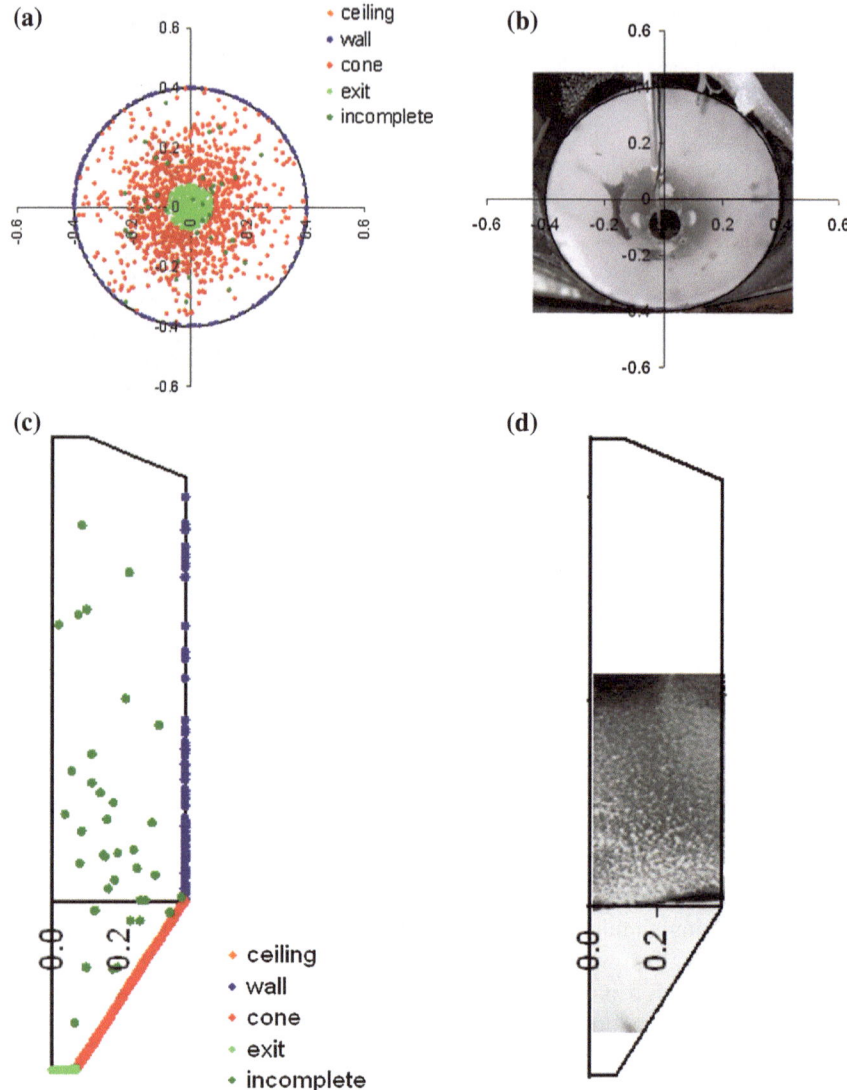

**Fig. 7.4** CFD-simulated (*left*) and experimental observations (*right*) of particle impact position on the *cone* (**a, b**) and side (**c, d**) walls (Anandharamakrishnan et al. 2010b)

Jet impingement is one of the techniques that utilizes energy efficiently and reduces processing time.

In the jet impingement technique, high velocity jets of fluid are injected through a nozzle onto a particular surface for the purpose of rapid heat and mass transfer. When a food product is heated or cooled using a fluid medium like air or water, a boundary layer forms. This boundary layer formation offers high resistance to heat transfer and makes the heating or cooling process slow. Therefore, the heating process becomes more time and energy consuming. However, this boundary layer problem can be solved using jet impingement, as jet impingements have both high velocity and turbulent flow. The high velocity of jets reduces the boundary layer between the heating surfaces and heating medium; hence, turbulent flow increases the heat transfer in food products (Jambunathan et al. 1992). Jet impingement heating is superior to conventional heating due to its advantages, such as higher efficiency, high heat and mass transfer rate, rapid moisture removal and uniform heating (Ovadia and Walker 1998; Li and Walker 1996; Nitin 2009).

For optimization of the jet impingement process, it is very important to understand its flow pattern, jet velocity profile, temperature distribution and effect of different parameters on heat transfer rate. The key parameters for effective heat transfer in jet impingement are nozzle geometry, nozzle to surface distance, velocity and temperature of jet (Garimella and Nenaydykh 1996). Some of the studies carried out by various researchers focused primarily on the effect of nozzle geometry on heat transfer (Garimella and Nenaydykh 1996, Zhao et al. 2004), the effects of single and multiplicity jets on heat transfer (Lou et al. 2005), the effects of pulsating jets on heat transfer (Zulkifli et al. 2009; Kurnia et al. 2012), etc. However, there is less understanding about the flow pattern of a jet. The effect of surrounding fluid, jet velocity and product position on the flow pattern of a jet needs to be explored. Moreover, it is very difficult to experimentally calculate the heat transfer coefficient at different positions (Kocer et al. 2007). Computational Fluid Dynamics (CFD) can be used to improve, understand and optimize the jet impingement process. CFD modeling of the jet impingement effect of nozzle geometry or position on heat transfer rate would be helpful in the analysis of effect of flow pattern on heat transfer, velocity and temperature profile, etc.

## 7.2.1 Flow Pattern of Impinging Jet

Jet flow is classified into three regions: a free jet region, a stagnation region and a wall jet region. The free jet region can be further classified into three sub-regions, such as the potential core region, the developing flow region and the developed flow region, as shown in Fig. 7.5. As the fluid exits from nozzle, the potential core region starts; it is a very short region, and the velocity of jet is almost similar to nozzle exit velocity and has no influence of surrounding air. Further, surrounding air starts mixing with jet air, resulting in the formation of turbulent peaks. As the downstream distance of the jet increases, velocity decreases. The potential core, developing and developed regions altogether constitute a free jet region. Further, when fluid jet reaches near the wall, the radial distance of the jet increases rapidly and velocity decreases, and this region called the wall jet region. Finally, the stagnant region is formed due to decrease in jet velocity to zero (Nitin 2009; Gardon and Akfirat 1965).

**Fig. 7.5**   Flow pattern in jet

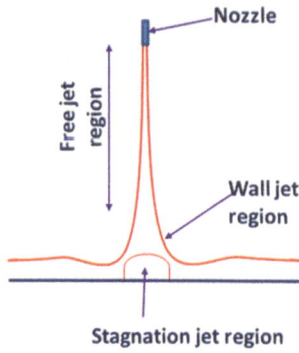

## 7.2.2  Effect of Nozzle Geometry on Heat Transfer

Selecting the jet configuration is very important, since nozzle shape and its geometry directly affect the length of the potential core region and the rate of energy dissipation (Jambunathan et al. 1992). Gariella and Nenaydykh (1996) reported that as nozzle to heating surface spacing increases, the effect of nozzle aspect ratio on the heat transfer coefficient is almost negligible. Compared to the aspect ratio, axial length to diameter of nozzle (X/D) and radial length to diameter of nozzle ratios (H/D) show much effect on the heat transfer coefficient (Sibulkin 1952).

Angioletti et al. (2005) studied CFD modeling of jet impingement at different Reynolds number values using three different turbulent models (Reynolds Normalized Group $k$-$\varepsilon$, $k$-$\omega$ Shear Stress Transport and Reynolds Stress Model), and validated with experimental data. Reynolds number values were between 1,000 and 4,000, and three models were compared with Particle Image Velocimetry (PIV) flow field at Re-1000 and h/d-4.5. Similarities between the results of the CFD ($k$–$\omega$) model and PIV can be observed.

Recently, Ma et al. (2012) developed a CFD model for a counter-current reactor and a confined jet mixer of a continuous hydrothermal flow synthesis (CHFS) system. They studied the effect of different operating conditions and reactor geometry on the flow field, temperature and concentration. They used the species transport equation coupled with the turbulent model k-$\varepsilon$. Velocity and temperature distribution were analyzed for the supercritical exit region.

Jafari and Alavi (2008) investigated the effect of various parameters on freezing time of slab-shaped food product, using air jet impingement with the help of CFD modeling. The RSM turbulent model was used for the modeling of turbulent flow of jet. In this study, the authors analyzed the effect of jet air temperature on temperature distribution, and jet air velocity on the flow and velocity field of the system. Reduction in freezing time was observed with the lowering of air temperature. However, as nozzle-to-surface spacing decreases, freezing time increases. Recently, Kurnia et al. (2012) analyzed the effect of pulsating impinging-jet on drying performance. Velocity vectors in the drying chamber for steady laminar and turbulent jets are shown in Fig. 7.6 (Kurnia et al. 2012). The RSM turbulent model was used for simulating turbulent flow

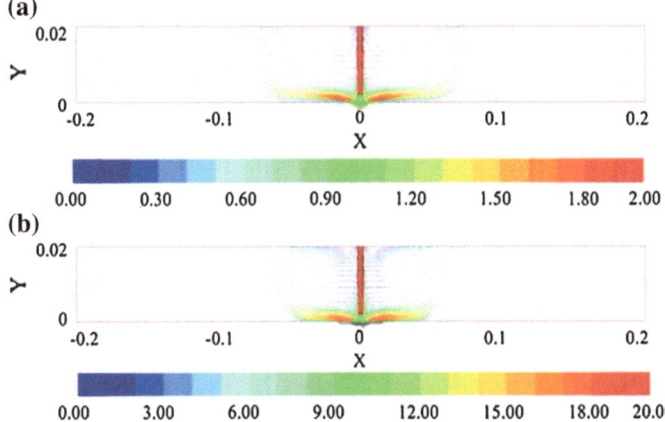

**Fig. 7.6** Velocity vectors in the drying chamber for steady laminar (**a**) and turbulent (**b**) jets (Kurnia et al. 2012)

of the impinging-jet dryer. The authors studied moisture content distribution, and their results showed uniform moisture distribution for a thin slab after 30 min of drying.

## 7.3 Application of CFD Modeling in the Flour Milling Industry

CFD modeling has a huge potential to address the problems faced by large scale grain processing industries. It is possible to model unit operations in the flour milling industry, where problems in handling and operating the equipment are encountered. CFD is also used in optimizing the processing conditions, so that losses in time, raw material and energy can be minimized. This increases the efficiency of the machine and decreases the economy (cost involved) in producing the desired product. However, the amount of research employed in this perspective is very limited.

A spouted bed dryer (an alternative to fluidized bed dryer) is used in drying coarse particles (such as grain) with narrower particle size distribution. Generally, three types of spouted beds shapes are used: (1) cylindrical (2) conical cylindrical, and (3) slot-rectangular (Cui and Grace 2008). A spouted bed dryer consists of injecting a high-speed gas–fluid that moves through a bed of solids, pushing and forcing the solids to the center of the container until reaching the upper level of the bed, where the solids later fall—through the effects of gravity—as a rain of particles forming a spout. One of the drawbacks with conventional spouted beds is the need for control of particle trajectory (Claflin and Fane 1984). However, CFD models can be used to solve this problem, and CFD models for spouted bed dryers can provide important information on the flow field within the spouted beds. This information can be utilized for process design, scale-up, optimization, and importantly, for reducing the need for experimentations.

A CFD model for the cylindrical spouted bed with spherical particles was developed using the Eulerian–Eulerian two-fluid modeling approach to study the

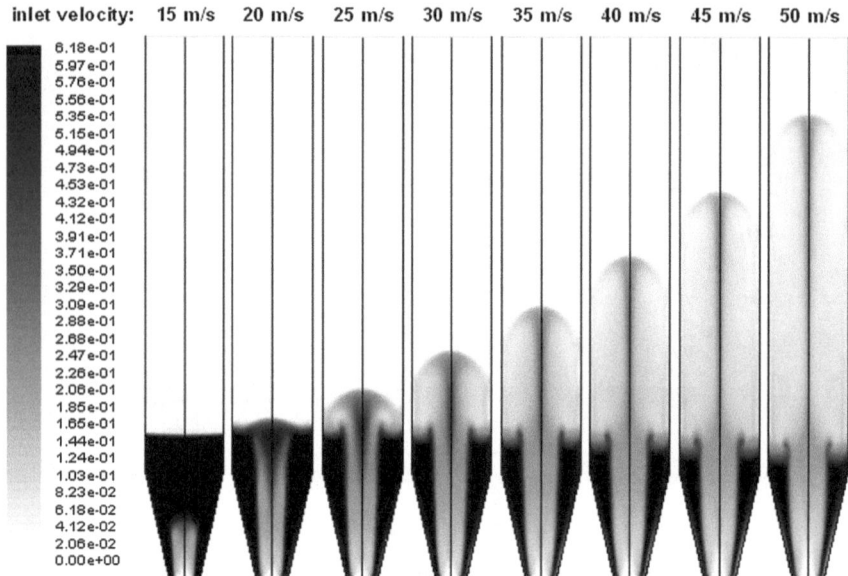

**Fig. 7.7**  Volume fraction distribution of grain for different air inlet velocities (Sobieski 2008)

gas-particle flow behavior (Zhonghua and Mujumdar 2008). Heat and mass transfer during grain drying in a spouted bed dryer was studied using a CFD (Szafran and Kmiec 2004). The Eulerian–Eulerian multiphase approach was used to predict gas–solid flow behavior. This model predicts mass transfer very well, but underpredicts heat transfer. Further, this CFD model was extended to study spouted-bed dryer hydrodynamics (Szafran et al. 2005). This model explains the interactions between the jet, fountain, and the annulus region and two phase flow of gas–solid very well, while traditional techniques fail to explain these interactions.

Flow behavior of grain, phase circulation and cluster formation in the loading region were predicted to be analogous to experimental observations. Sobieski (2008) used the Eulerian multiphase approach to model grain drying in a spouted bed dryer. CFD model predictions of volume fraction distribution of grain for various inlet air velocities are shown in Fig. 7.7 (Sobieski 2008). In this study, Sobieski found that inlet air velocities play a crucial role in the behaviour of particles inside the spouted bed dryer. Figure 7.8 shows the distribution of volume fraction for various turbulence models (Sobieski 2008). Each turbulence model predicts a different fountain height. The effect of grain density, diameter, change in volume and packing coefficient on the distribution of the grain height was studied by Sobieski (2008).

Recently, Sobieski (2010) discussed the flow analysis inside the spouted bed dryer in terms of creation of geometry; selection of computational space and grid; the Eulerian multiphase model; and the turbulence model and its sensitivity to phase changes and flow parameters. Further, spout, fountain and annular regions were well predicted by the above model (Fig. 7.9). This study was performed using an Eulerian multiphase model.

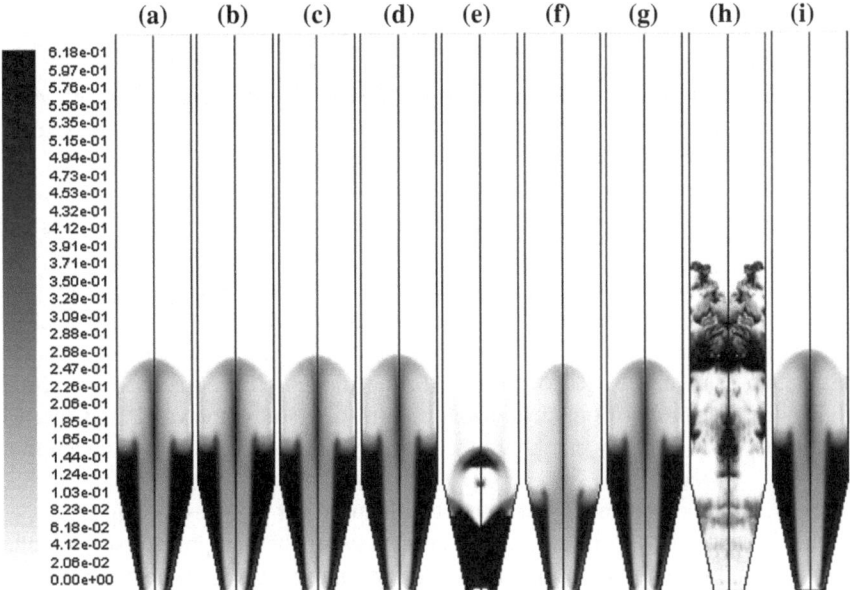

**Fig. 7.8** Distribution of volume fraction for different turbulence models (inlet velocity of 30 m = s); (**a**) standard $\kappa$-$\varepsilon$ model with standard wall function (SWF) (dispersed); (**b**) standard $\kappa$-$\varepsilon$ model with enhanced wall treatment (EWT) (dispersed); (**c**) $\kappa$-$\varepsilon$ RNG model with SWF (dispersed); (**d**) $\kappa$-$\varepsilon$ RNG model with EWT (dispersed); (**e**) $\kappa$-$\varepsilon$ realizable model with SWF (dispersed); (**f**) $\kappa$-$\varepsilon$ realizable model with EWT (dispersed); (**g**) Reynolds stress model with SWF (dispersed); (**h**) Reynolds stress model with EWT(dispersed); and (**i**) laminar flow (Sobieski 2008)

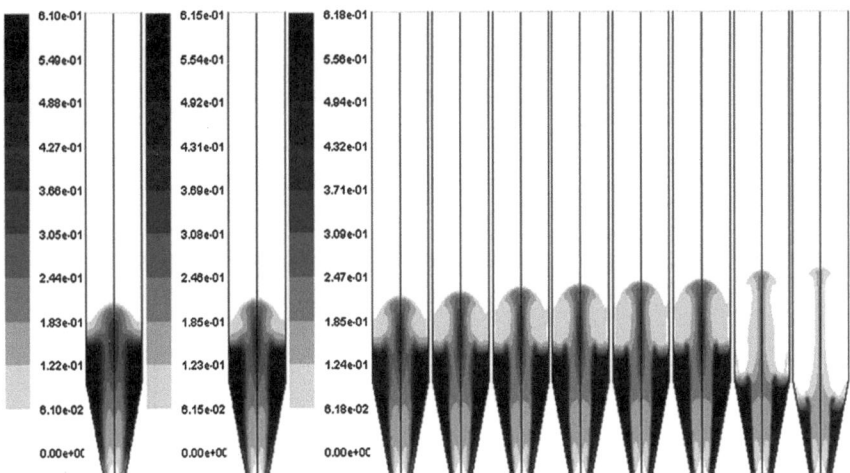

**Fig. 7.9** Volume fraction occupied by the granular phase for differently sized structural grids (from the *left*: 2,160; 4,284; 6,615; 11,732; 17,136; 22,347; 27,090; 32,900; 40,144; and 47,880) (Sobieski 2010)

## 7.4  CFD Modeling of Fumigation of Flour Mills

Flour mills normally use fumigation to kill insects that can create damage to the grain. The fumigant used was methyl bromide. This process can be structurally modeled using CFD, where the model will have two phases, such as external flow model and internal flow model. The external flow model includes the flour mill and surrounding structures to predict stagnation pressures on the mill's walls as a function of wind speed and direction data. The internal flow model includes interior details of the mill, such as building plans, and locations of major equipment, partitions, piping and ducting. The cracks are represented as an effective leakage zone (Chayaprasert et al. 2008). The locations of fumigant introduction sites and circulation fans are also accounted for in the model. Chayaprasert et al. (2008) concluded that the CFD models developed in their study are valid, and that the established methodology can be utilized for any type of structure for modeling of fumigation process.

Thus, CFD is a promising tool that can be used for troubleshooting unit operations. It can also help in optimizing conditions of the equipment used in roller flour mill industries. For optimizing operating conditions, experiments are usually performed on the equipment itself; this is time-consuming, laborious and expensive. However, these problems can be solved by using CFD to model the equipment (or system, rather than performing experiments on the system itself). Hence, CFD is a powerful and pervasive tool for process and product development in the food-processing sector.

# References

Abrahamsson B, Pal A, Sjoberg M, Carlsson M, Laurell E, Brasseur JG (2005) A novel in vitro and numerical analysis of shear-induced drug release from extended-release tablets in the fed stomach. Pharm Res 22(8):1215–1226

Anandharamakrishnan C (2003) Computational fluid dynamics (CFD)—applications for the food industry. FRI 22(6):62–68

Anandharamakrishnan C, Khwanpruk K, Rielly CD, Stapley AGF (2006) Spray-freeze-drying at sub-atmospheric pressures. Drying 2006—proceedings of the 15th international drying symposium, vol B, Budapest, Hungary, Aug 20–23, pp 636–642

Anandharamakrishnan C, Rielly CD, Stapley AGF (2007) Effects of process variables on the denaturation of whey proteins during spray-drying. Dry Technol 25:799–807

Anandharamakrishnan C, Gimbun J, Stapley AGF, Rielly CD (2010) A study of particle histories during spray drying using computational fluid dynamic simulations. Dry Technol 28:566–576

Anandharamakrishnan C, Gimbun J, Stapley AGF, Rielly CD (2010) Application of computational fluid dynamic (CFD) simulations to spray-freezing operations. Dry Technol 28:94–102

Anandpaul D, Anishaparvin A, Anandharamakrishnan C (2011) Computational fluid dynamics studies on pasteurisation of canned milk. Int J Dairy Technol 64(2):305–313

Anderson JD (1984) Computational fluid dynamics—the basics with applications. McGraw-Hill Inc, New York

Angioletti M, Nino E, Ruocco G (2005) CFD turbulent modelling of jet impingement and its validation by particle image velocimetry and mass transfer measurements. Int J Therm Sci 44:349–356

Anishaparvin A, Chhanwal N, Indrani D, Raghavarao KSMS, Anandharamakrishnan C (2010) An investigation of bread baking process in a pilot-scale electrical heating oven using computational fluid dynamics. J Food Sci 75:E605–E611

Augusto PED, Cristianini M (2010a) Evaluation of geometric symmetry condition in numerical simulations of thermal process of packed liquid food by computational fluid dynamics (CFD). Int J Food Eng 6(5)

Augusto PED, Cristianini M (2010b) Computational fluid dynamics analysis of viscosity influence on thermal in-package liquid food process. Int J Food Eng 6(6)

Augusto PED, Cristianini, M (2011) Numerical simulation of packed liquid food thermal process using computational fluid dynamics (CFD). Int J Food Eng 7 (4)

Augusto PED, Pinherio TF, Cristianini M (2010) Using computational fluid dynamics (CFD) for the evaluation of beer pasteurization: effect of orientation of cans. Ciencia e Tecnologia de Alimentos 30(4):980–986

Balimane PV, Chong S (2005) Cell culture-based modules for intestinal permeability: a critique. Drug Discov Today 10(5):335–343

C. Anandharamakrishnan, *Computational Fluid Dynamics Applications in Food Processing*, SpringerBriefs in Food, Health, and Nutrition, DOI: 10.1007/978-1-4614-7990-1, © Chinnaswamy Anandharamakrishnan 2013

Birchal VS, Huang L, Mujumdar AS, Passos ML (2006) Spray dryers- modeling and simulation. Dry Technol 24:359–371

Bird B, Stewart WE, Lightfoot EN (1960) Transport phenomena. Wiley, New York

Boulet M, Marcos B, Dostie M, Moresoli C (2010) CFD modeling of heat transfer and flow field in a bakery pilot oven. J Food Eng 97:393–402

Camilleri M, Prather CM (1993) Gastric motor physiology and motor disorders. In: Feldmann M, Scharschmidt BF, Sleisenger MH (eds) Gastrointestinal and liver disease: pathophysiology/ diagnosis/management, 6th edn. WB Saunders Co., Philadelphia, pp 572–586

Cauvain SP (2003) Bread making-improving quality. CRC Press, New York

Charm SE (1971) The fundamentals of food engineering. The AVI Publishing Company, Westport

Chayaprasert W, Maier DE, Ileleji KE, Murthy JY (2008) Development and validation of computational fluid dynamics models for precision structural fumigation. J Stored Prod Res 44:11–20

Chen XD, Xie GZ (1997) Fingerprints of the drying behavior of particulate or thin layer food materials established using a reaction engineering model. Trans Inst Chem Eng, Part C 75:213–222

Chen XD, Jin Y (2009) Numerical study of the drying process of different sized particles in an industrial-scale spray dryer. Dry Technol 27:371–381

Chen XD, Jin Y (2009) A three-dimensional numerical study of the gas/particle interactions in an industrial-scale spray dryer for milk powder production. Dry Technol 27:1018–1027

Chen XD, Pirini W, Ozilgen M (2001) The reaction engineering approach to modeling drying of thin layer of pulped kiwi fruit flesh under conditions of small Biot numbers. Chem Eng Prog 40:165–181

Chhanwal N, Anishaparvin A, Indrani D, Raghavarao KSMS, Anandharamakrishnan C (2010) Computational fluid dynamics (CFD) modeling of an electrical heating oven for bread-baking process. J Food Eng 100:452–460

Chhanwal N, Indrani D, Raghavarao KSMS, Anandharamakrishnan C (2011) Computational fluid dynamics modeling of bread baking process. Food Res Int 44:978–983

Chhanwal N, Tank A, Raghavarao KSMS, Anandharamakrishnan C (2012) Computational fluid dynamics (CFD) modeling for bread baking process-a review. Food Bioprocess Tech 5:1157–1172

Claflin JK, Fane AG (1984) Fluid mechanics, heat transfer in spouted beds with draft tubes. Dry 137–141

Claussen C, Ustad TS, Strommen I, Walde PM (2007) Atmospheric freeze drying—a review. Dry Technol 25:957–967

Cortella G, Manzan M, Comini G (1998) Computation of air velocity and temperature distributions in open display cabinets. In: Advances in the refrigeration systems, food technologies and cold chain. International Institute of Refrigeration, Paris

Crowe CT, Sharam MP, Stock DE (1977) The particle source in cell (PSI-Cell) model for gas-droplet flows. J Fluid Eng 9:325–332

Cui H, Grace JR (2008) Spouting of biomass particles: a review. Bioresource Technol 99:4008–4020

Datta AK, Teixeira AA (1987) Numerical modelling of natural convection heating in canned liquid foods. Trans ASAE 30(5):1542–1551

Davey LM, Pham QT (1997) Predicting the dynamic product heat load and weight loss during beef chilling using a multi-region finite difference approach. Int J Refrig 20(7):470–482

Davey LM, Pham QT (2000) A multi-layered two-dimensional finite element model to calculate dynamic product heat load and weight loss during beef chilling. Int J Refrig 23(6):444–456

Denys S, Pieters JG, Dewettinck K (2003) Combined CFD and experimental approach for determination of the surface heat transfer coefficient during thermal processing of eggs. J Food Sc 68:943–951

Denys S, Pieters JG, Dewettinck K (2004) Computational fluid dynamics analysis of combined conductive and convective heat transfer in model eggs. J Food Eng 63:281–290

Denys S, Dewettinck K, Pieters JG (2005) CFD analysis for process impact assessment during thermal pasteurization of intact eggs. J Food Protect 68:366–374

DeSesso JM, Jacobson CF (2001) Anatomical and physiological parameters affecting gastrointestinal absorption in humans and rats. Food Chem Toxicol 39:209–228

DeVries U, Velthuis H, Koster K (1994) Baking ovens and product quality: a computer model. Food Sci Technol Today 9(4):232–234

Dhall A, Datta AK, Torrance KE, Almeida MF (2009) Radiative heat exchange modeling inside an oven. AIChE J 55:2448–2460

Dimou A, Yanniotis S (2011) 3D numerical simulation of asparagus sterilization in a still-can using computational fluid dynamics. J Food Eng 104:394–403

Dokoumetzidis A, Valsami G, Macheras P (2007) Modelling and simulation in drug absorption processes. Xenobiotica 37(10–11):1052–1065

Ducept F, Sionneau M, Vasseur J (2002) Superheated steam dryer—simulations and experiments on product drying. Chem Eng J 86:75–83

Einhorn M (1898) Anatomy and physiology. In: Diseases of the stomach: a text-book for practitioners and students. William Wood and Company, New York

Erdogdu F, Tutar M (2010) Velocity and temperature field characteristics of water and air during natural convection heating in cans. J Food Sci 76(1):119–129

Farid M, Ghani AG (2004) A new computational technique for the estimation of sterilization time in canned food. Chem Eng Process 43:523–531

Fellows PJ (1998) Food processing technology, 2nd edn. CRC Press, New York

Ferrua MJ, Singh RP (2010) Modeling the fluid dynamics in a human stomach to gain insight of food digestion. J Food Sci 75(7):151–161

Ferrua MJ, Kong F, Singh RP (2011) Computational modelling of gastric digestion and the role of food material properties. Trends Food Sci Tech 1–12

Fletcher AJ (2000) Computational techniques for fluid dynamics, 2nd edn. Springer, New York

Fluent (2006) Fluent user's guide. Ansys Inc, USA

Gabites JR, Abrahamson J, Winchester JA (2010) Air flow patterns in an industrial milk powder spray dryer. Chem Eng Res Des 88(7):899–910

Gardon R, Akfirat JC (1965) Role of turbulence in determining the heat transfer characteristics of jet impingement. Int J Heat Mass Tran 8:1261–1272

Garimella SV, Nenaydykh B (1996) Nozzle-geometry effects in liquid jet impingement heat transfer. Int J Heat Mass Tran 39:2915–2923

Geliebter A, Mellon PM, McCray RS, Gallagher DR, Gage D, Hashim SA (1992) Gastric capacity, gastric emptying, and test-meal intake in normal and bulimic women. Am J Clin Nutr 56(4):656–661

Ghani AG, Farid MM, Chen XD, Richards P (1999) Numerical simulation of natural convection heating of canned food by computational fluid dynamics. J Food Eng 41(1):55–64

Ghani AG, Farid MM, Chen XD, Richards P (1999) An investigation of deactivation of bacteria in a canned liquid food during sterilisation using computational fluid dynamics (CFD). J Food Eng 42(4):207–214

Ghani AG, Farid MM, Che XD, Richards P (2001) Thermal sterilization of canned food in a 3-D pouch using computational fluid dynamics. J Food Eng 48:147–156

Ghani AG, Farid MM, Chen XD (2002) Numerical simulation of transient temperature and velocity profiles in a horizontal can during sterilization using computational fluid dynamics. J Food Eng 51:77–83

Ghani AG, Farid MM, Chen XD (2002) Theoretical and experimental investigation of the thermal inactivation of *Bacillus stearothermophilus* in food pouches. J Food Eng 51:221–228

Ghani AG, Farid MM, Chen XD (2002) Theoretical and experimental investigation of the thermal destruction of Vitamin C in food pouches. Comput Electron Agric 34:129–143

Ghani AG, Farid MM, Chen XD (2003) The effect of can rotation on the thermal sterilization of liquid food using computational fluid dynamics. J Food Eng 57:9–16

Ghani AG, Farid MM (2006) Using the computational fluid dynamics to analyze the thermal sterilization of solid-liquid food mixture in cans. Innov Food Sci Emerg 7:55–61

Ghani AGA, Farid MM (2007a) Modeling of high pressure food processing using CFD. In: Sun DW (ed) Computational fluid dynamics in food processing. CRC Press, Boca Raton

Ghani AGA, Farid MM (2007) Numerical simulation of solid-liquid food mixture in a high pressure processing unit using computational fluid dynamics. J Food Eng 80:1031–1042

Goula AM, Adamopoulos KG (2004) Influence of spray drying conditions on residue accumulation-simulation using CFD. Dry Technol 22:1107–1128

Guyton AC, Hall JE (2006) Textbook of medical physiology. Elsevier, Philadelphia

Heldman DR (1974) An analysis of atmospheric freeze drying. J Food Sci 39:147–155

Hiiemae K, Palmer JB (1999) Food transport and bolus formation during complete feeding sequences on foods of different initial consistency. Dysphagia 14:31–42

Hindmarsh JP, Russell AB, Chen XD (2003) Experimental and numerical analysis of the temperature transition of a suspended freezing water droplet. Int J Heat Mass Tran 46:1199–1213

Holdsworth D, Simpson R (2007) Thermal processing of packaged foods. Food engineering series, 2nd edn. Springer, New York

Hu Z, Sun DW (1999) The temperature distribution of cooked meat joints in an air-blast chiller during cooling process: CFD simulation and experimental verification. Paper presented at the 20th international congress of refrigeration, Sydney

Hu Z, Sun DW (2000) Simulation of heat and mass transfer for vacuum cooling of cooked meats by using computational fluid dynamics code. Paper presented at the 8th international congress on engineering and food, Puebla

Huang LX, Kumar K, Mujumdar AS (2003) Use of computational fluid dynamics to evaluate alternative spray chamber configurations. Dry Technol 21:385–412

Huang LX, Kumar K, Mujumdar AS (2003) A parametric study of the gas flow patterns and drying performance of co-current spray dryer: results of a computational fluid dynamics study. Dry Technol 21(6):957–978

Huang LX, Kumar K, Mujumdar AS (2004) Simulation of a spray dryer fitted with a rotary disk atomizer using a three-dimensional computational fluid dynamic model. Dry Technol 22(6):1489–1515

Huang LX, Passos ML, Kumar K, Mujumdar AS (2005) A three dimensional simulation of a spray dryer fitted with a rotary atomizer. Dry Technol 23:1859–1873

Huang LX, Kumar K, Mujumdar AS (2006) A comparative study of a spray dryer with rotary disc atomizer and pressure nozzle using computational fluid dynamic simulations. Chem Eng Process 45:461–470

Huang LX, Mujumdar AS (2007) Simulation of an industrial spray dryer and prediction of off-design performance. Dry Technol 25:703–714

Jafari M, Alavi P (2008) Analysis of food freezing by slot jet impingement. J Appl Sci 8(7):1188–1196

Jakobsen HA, Sannaes BH, Grevskott S, Svendsen HF (1997) Modelling of vertical bubble-driven flows. Ind Eng Chem Res 36:4052–4074

Jambunathan K, Lai E, Moss MA, Button BL (1992) A review of heat transfer data for single circular jet impingement. Int J Heat Fluid Fl 13(2):106–15

Juliano P, Knoerzer K, Fryer P, Versteeg C (2009) *C. botulinum* inactivation kinetics implemented in a computational model of a high pressure sterilization process. Biotechnol Prog 25:163–175

Kannan A, Sandaka PCHG (2008) Heat transfer analysis of canned food sterilization in still retort. J Food Eng 88:213–228

Keet AD (1993) Infantile hypertrophic pyloric stenosis. In: The pyloric sphincteric cylinder in health and disease. Springer, Berlin

Khurana M, Karwe MV (2009) Numerical prediction of temperature distribution and measurement of temperature in a high hydrostatic pressure food processor. Food Bioprocess Tech 2:279–290

Kieviet FG, Kerkhof PJAM (1995) Measurements of particle residence time distributions in a co-current spray dryer. Dry Tech 13(5–7):1241–1248

Kieviet, FG, Kerkhof PJAM (1996) In: Mujumdar AS (ed) Using computational fluid dynamics to model product quality in spray drying: air flow temperature and humidity patterns. In Drying'96, vol A, Krakow, pp 259–266

Kieviet FG (1997) Modeling quality in spray drying. Ph.D. thesis, Endinhoven University of Technology, Netherlands

Kim DY, Camilleri M, Murray JA, Stephens DA, Levine JA, Burton DD (2001) Is there a role for gastric accommodation and satiety in asymptomatic obese people? Obesity Res 9(11):655–661

Kiziltas S, Erdogdu F, Palazoglu TK (2010) Simulation of heat transfer for solid–liquid food mixtures in cans and model validation under pasteurization conditions. J Food Eng 97:449–456

Knoerzer K, Juliano P, Gladman S, Versteeg C, Fryer PJ (2007) A computational model for temperature and sterility distributions in a pilot-scale high-pressure high-temperature process. AIChE J 53:2996–3010

Knoerzer K, Buckow R, Juliano P, Chapman B, Versteeg C (2010) Carrier optimisation in a pilot-scale high pressure sterilization plant—an iterative CFD approach. J Food Eng 97:199–207

Knoerzer K, Chapman B (2011) Effect of material properties and processing conditions on the prediction accuracy of a CFD model for simulating high pressure thermal (HPT) processing. J Food Eng 104:404–413

Kocer D, Nitin N, Karwe MV (2007) Application of CFD in jet impingement oven. In: Sun DW (ed) Computational fluid dynamics in food processing. CRC Press, Boca Raton, pp 469–485

Kong F, Singh RP (2008) A model stomach system to investigate disintegration kinetics of solid foods during gastric digestion. J Food Sci 73(5):210

Kong F, Singh RP (2010) A human gastric simulator (hgs) to study food digestion in human stomach. J Food Sci 75(9):E627–E635

Koribilli N, Aravamudan K, Varadhan A (2011) Quantifying enhancement in heat transfer due to natural convection during canned food thermal sterilization in a still retort. Food Bioprocess Tech 4:429–450

Koutchma T (2012) Design for high-pressure processing. In: Ahmed J, Rahman MS (eds) Handbook of food process design. Wiley-Blackwell, Oxford, pp 998–1030

Kumar A, Bhattacharya M, Blaylock J (1990) Numerical simulation of natural convection heating of canned thick viscous liquid food products. J Food Sci 55(5):1403

Kumar A, Bhattacharya M (1991) Transient temperature and velocity profiles in a canned Non-Newtonian liquid food during sterilization in a still-cook retort. Int J Heat Mass Tran 34(4–5):1083–1096

Kuo KKY (1986) Principles of combustion. Wiley, New York

Kuriakose R, Anandharamakrishnan C (2010) Computational fluid dynamics applications in spray drying of food products. Trends Food Sci Tech 21:383–398

Kurnia JC, Sasmito AP, Tong W, Mujumdar AS (2012) Energy- efficient thermal drying using impinging-jets with time-varying heat input—a computational study. J Food Eng 114:269–277

Langrish TAG, Oakley DE, Keey RB, Bahu RE, Hutchinson CA (1993) Time-dependent flow patterns in spray dryers. Trans Inst Chem Eng Part A 71:355–360

Langrish TAG, Zbicinski I (1994) The effects of air inlet geometry and spray cone angle on the wall deposition rates in spray dryer. Trans Inst Chem Eng Part A 72:420–430

Langrish TAG, Fletcher DF (2001) Spray drying of food ingredients and applications of CFD in spray drying. Chem Eng Process 40:345–354

Langrish TAG, Fletcher DF (2003) Prospects for modelling and design of spray dryers in the 21st century. Dry Technol 21:197–215

Langrish TAG (2007) New engineered particles from spray dryers: research needs in spray drying. Dry Technol 25:981–993

Langrish TAG, Williams J, Fletcher DF (2004) Simulation of the effects of swirl on gas flow patterns in a pilot-scale spray dryer. Trans Inst Chem Eng Part A 82(A7):821–833

Launder BE, Spalding DB (1972) Lectures in mathematical models of turbulence. Academic Press, London

Leuenberger L, Plitzko M, Puchkov M (2006) Spray freeze drying in fluidized bed at normal and low pressure. Dry Technol 7:11–719

Li A, Walker CE (1996) Cake baking in conventional, impingement and hybrid ovens. J Food Sci 61(1):188–191

Liu L, Fishman ML, Kost J, Hicks KB (2003) Pectin-based systems for colon-specific drug delivery via oral route. Biomaterials 24:3333–3343

Lou ZQ, Mujumdar AS, Yap C (2005) Effects of geometric parameters on confined impinging jet heat transfer. Appl Therm Eng 25:2687–2697

Love RJ, Lentle RG, Asvarujanon P, Hemar Y, Stafford KJ (2012) An expanded finite element model of the intestinal mixing of digesta. Food Digest 1–10. doi: 10.1007/s13228-012-0017-x

Ma CY, Wang XZ, Tighe CJ Darr JA (2012) Modelling and simulation of counter-current and confined jet reactors for continuous hydrothermal flow synthesis of nano-materials. 8th IFAC symposium on advanced control of chemical processes Furama Riverfront, Singapore

MacLeod CS, McKittrick JA, Hindmarsh JP, Johns ML, Wilson DI (2006) Fundamentals of spray freezing of instant coffee. J Food Eng 74:451–461

Mahler GJ, Esch MB, Tako E, Southard TL, Archer SD, Glahn RP, Shuler ML (2012) Oral exposure to polystyrene nanoparticles affects iron absorption. Nat Nanotechnol 7(4):164–271

Malecki GJ, Shinde P, Morgan AI, Farkas DF (1970) Atmospheric fluidized bed freeze drying. Food Technol 24:601–603

Mariotti M, Rech G, Romagnoni P (1995) Numerical study of air distribution in a refrigerated room. Proceedings of international congress of refrigeration, vol 2. Den Hague, The Netherlands, pp 98–105

Marshall EM, Bakker A (2002) Computational fluid mixing. Fluent Inc, Lebanon

Masters K (1991) Spray drying. Wiley, Essex

Mayer EA (1994) The physiology of gastric storage and emptying. In: Johnson L (ed) Physiology of the gastrointestinal tract, 3rd edn. Raven Press, New York, pp 929–76

Mezhericher M, Levy A, Borde I (2008) Droplet-droplet interactions in spray drying by using 2D computational fluid dynamics. Dry Technol 26:265–282

Mezhericher M, Levy A, Borde I (2009) Modeling of droplet drying in spray chambers using 2D and 3D computational fluid dynamics. Dry Technol 27:359–370

Mills D (1998–1999) Development and validation of a preliminary model for optimisation of baking ovens. The Food and Packaging Cooperative Research Centre Annual Report (1998-1999), Australia

Mistry H, Ganapathi-subbu S, Dey S, Bishnoi P, Castillo JL (2006) Modeling of transient natural convection heat transfer in electric ovens. Appl Therm Eng 26:2448–2456

Morsi SA, Alexander AJ (1972) An investigation of particle trajectories in two-phase flow systems. J Fluid Mech 55(2):193–208

Mostafa AA, Mongia HC (1987) On the modeling of turbulent evaporating sprays: Eulerian versus Lagrangian approach. Int J Heat Mass Tran 30(12):2583–2593

Moureh J, Derens E (2000) Numerical modelling of the temperature increase in frozen food packaged in pallets in the distribution chain. Int J Refrig 23(7):540–552

Nijdam JJ, Guo B, Fletcher DF, Langrish TAG (2004) Challenges of simulating droplet coalescence within a spray. Dry Technol 22(6):1463–1488

Nijdam JJ, Guo B, Fletcher DF, Langrish TAG (2006) Lagrangian and Eulerian models for simulating turbulent dispersion and coalescence of droplets within a spray. Appl Math Model 30:1196–1211

Nitin N (2009) A dissertation on transport phenomenon in jet impingement baking. Ph.D Thesis, Graduate School-New Brunswick Rutgers, The State University of New Jersey

Norton IT, Fryer PJ, Moore S (2006) Product/process integration in food manufacture: engineering sustained health. AIChE J 52(5):1632–1640

Norton T, Sun DW (2006) Computational fluid dynamics (CFD)—an effective and efficient design and analysis tool for the food industry: a review. Trends Food Sci Tech 17:600–620

Oakley DE, Bahu RE (1993) Computational modelling of spray dryers. Comp Chem Eng 17:493–498

Ovadia DZ, Walker CE (1998) Impingement in food processing. Food Technol 52(4):46–50

Padmavathi R, Anandharamakrishnan C (2012) Computational fluid dynamics modeling of the thermal processing of canned pineapple slices and titbits. Food Bioprocess Tech 6(4):882–895

Pal A, Indireshkumar K, Schwizer W, Abrahamsson B, Fried M, Brasseur JG (2004) Gastric flow and mixing studied using computer simulation. Proc Roy Soc London Biol Sci 271:2587–2594

Pal A, James GB, Abrahamsson B (2007) A stomach road or 'magenstrasse' for gastric emptying. J Biomech 40:1202–1210

Pallotta N, Cicala M, Frandina C, Corazziari E (1998) Antropyloric contractile patterns and transpyloric flow after meal ingestion in humans. Am J Gastroenterol 93:2513–2522

Papadakis SE, King CJ (1988) Air temperature and humidity profiles in spray drying. 1. Features predicted by the particle source in cell model. Ind Eng Chem Res 27:2111–2116

Parrott N, Lave T (2010) Computer models for predicting drug absorption. In: Dressman JB, Reppas C (eds) Oral drug absorption: prediction and assessment. Informa Healthcare, New York, pp 338–355

Paton J, Khatir Z, Thompson H, Kapur N, Toropov V (2012) Thermal energy management in the bread baking industry using a system modeling approach. Appl Therm Eng. doi:10.1016/j.applthermaleng.2012.03,036

Perry RH, Chilton CH (1984) Perry's chemical engineers handbook. McGraw-Hill, London

Purlis E, Salvadori VO (2009) Bread baking as a moving boundary problem. Part 2: model validation and numerical simulation. J Food Eng 91:434–442

Rabiey L, Flick D, Duquenoy A (2007) 3D simulations of heat transfer and liquid flow during sterilization of large particles in a cylindrical vertical can. J Food Eng 82:409–417

Rao MA, Anantheswaran RC (1988) Convective heat transfer to fluid foods in cans. Adv Food Res 32:39–84

Reay D (1988) Fluid flow, residence time simulation and energy efficiency in industrial dryers, In: Roques M (ed) Proceedings of sixth international drying symposium IDS, Versailles, France, pp KL1–KL8

Roache PJ (1976) Computational fluid dynamics. Hermosa Publishers, Albuquerque

Roustapour OR, Hosseinalipour M, Ghobadian B, Mohaghegh F, Azad NM (2009) A proposed numerical-experimental method for drying kinetics in a spray dryer. J Food Eng 90:20–26

Sahu AK, Kumar P, Patwardhan AW, Joshi JB (1999) CFD modelling and mixing in stirred tanks. Chem Eng Sci 54(13–14):2285–2293

Schulze K, Herman RJ, Shirazi SS, Brown BP, Lammers WJEP, Donck LV, Stephen B, Smets D, Schuurkes JAJ, Shirazi S (1998) Contractions move contents by changing the configuration of the isolated cat stomach. Am J Physiol Gastrointest Liver Physiol 274:G359–G369

Scott GM (1977) Simulation of the flow of non-newtonian foods using computational fluid dynamics. Campden & Chorleywood Food Research Association R & D Report No. 34, Chipping Campden

Scott G, Richardson P (1997) The applications of computational fluid dynamics in the food industry. Trends Food Sci Tech 8:119–124

Sibulkin M (1952) Heat transfer near the forward stagnation point of a body of revolution. J Aeronaut Sci 19:570–571

Siriwattanayotin S, Yoovidhya T, Meepadung T, Ruenglertpanyakul M (2006) Simulation of sterilization of canned liquid food using sucrose degradation as an indicator. J Food Eng 73:307–312

Smith AC (2004) Texture and mastication. Texture in food, vol 2, Solid foods. Woodhead Publishing Ltd, Cambridge

Sobieski W (2008) Numerical analysis of sensitivity of Eulerian multiphase model for a spouted-bed grain dryer. Dry Technol 26(12):1438–1456

Sobieski W (2010) Selected aspects of developing a simulation model of a spouted bed grain dryer based on the Eulerian multiphase model. Dry Technol 28:1331–1343

Southwell DB, Langrish TAG, Fletcher DF (1999) Process intensification in spray dryer by turbulent enhancement. Trans Inst Chem Eng Part A 77(3):189–205

Sparrow EM, Abraham JP (2003) A computational analysis of the radiative and convective process that take place in preheated and non-preheated ovens. Heat Transfer Eng 24:25–37

Spratt P, Nicolella C, Pyle DL (2005) An Engineering Model of the Human Colon. Trans IChemE C 83:147–157

Stoll BR, Richard PB, Leipold HR, Milstein S, Edwards DA (2000) A theory of molecular absorption from the small intestine. Chem Eng Sci 55:473–489

Straatsma J, Van Houwelingen G, Steenbergen AE, De Jong P (1999) Spray drying of food products: 1. Simulation model. J Food Eng 42(2):67–72

Sun DW (2007) Computational fluid dynamics in food processing. CRC Press, Taylor and Francis group, Boca Raton

Szafran G, Kmiec A (2004) CFD modeling of heat and mass transfer in a spouted bed dryer. Ind Eng Chem Res 43:1113–1124

Szafran G, Kmiec A, Ludwig W (2005) CFD modeling of a spouted-bed dryer hydrodynamics. Dry Technol 23:1723–1736

Tank A, Chhanwal N, Indrani D, Anandharamakrishnan C (2012) Computational fluid dynamics modeling of bun baking process under different oven load conditions. J Food Sci Tech. doi:10.1007/s13197-012-0736-6

Tannehill JC, Anderson DA, Pletcher RH (1997) Computational fluid mechanics and heat transfer, 2nd edn. Taylor & Francis, Philadelphia

Tattiyakul J, Rao MA, Datta AK (2002) Heat transfer to a canned corn starch dispersion under intermittent agitation. J Food Eng 54:321–329

Tharakan A, Rayment P, Fryer PJ, Norton IT (2007) Modelling of physical and chemical processes in the small intestine. Proceedings of European congress of chemical engineering (ECCE-6), Copenhagen, pp 16–20

Therdthai N, Zhou W (2003) Recent advances in the studies of bread baking process and their impact on the bread baking technology. Food Sci Tech Res 9(3):219–226

Therdthai N, Zhou W, Adamczak T (2003) Two dimensional CFD modeling and simulation of an industrial continuous bread baking oven. J Food Eng 60:211–217

Therdthai N, Zhou W, Adamczak T (2004) Simulation of starch gelatinization during baking in a traveling—tray oven by integrating a three dimensional CFD model with a kinetic model. J Food Eng 65:543–550

Therdthai N, Zhou W, Adamezak T (2004) Three-dimensional CFD modeling and simulation of the temperature profiles and airflow patterns during a continuous industrial baking process. J Food Eng 65:599–608

Ullum T (2006) Simulation of a spray dryer with a rotary atomizer: the appearance of vortex breakdown. In: Farkas I, Mujumdar AS (eds) Drying 2006—proceedings of the 15th international drying symposium. Budapest, Hungary, pp 251–257

Vanina FM, Lucasa T, Trystram G (2009) Crust formation and its role during bread baking. Trends Food Sci Tech 20:333–343

Varma MN, Kannan A (2005) Enhanced food sterilization through inclination of the container walls and geometry modifications. Int J Heat Mass Tran 48(18):3753–3762

Varma MN, Kannan A (2006) CFD studies on natural convective heating of canned food in conical and cylindrical containers. J Food Eng 77:1024–1036

Verboven P, Nicolai B, Delele M, Schenk A, Ramon H (2009) Evaluation of a chicory root cold store humidification system using computational fluid dynamics. J Food Eng 94(1):110–121

Versteeg HK, Malalasekera W (1995) An introduction to computational fluid dynamics. Pearson Education Ltd, Essex, England

Verboven P, Scheerlinck N, de Baerdemaeker J, Nicolai BM (2000) Computational fluid dynamics modelling and validation of the isothermal airflow in a forced convection oven. J Food Eng 43:41–53

Verboven P, Nicolai B, Delele M, Tijskens E, Atalay Y, Ho Q, Ramon H (2008) Combined discrete element and CFD modelling of airflow through random stacking of horticultural products in vented boxes. J Food Eng 89(1):33–41

Wagner MJ, Lucas T, Le Ray D, Trystram G (2007) Water transport in bread during baking. J Food Eng 78:1167–1173

Wang L, Sun DW (2003) Recent developments in numerical modelling of heating and cooling processes in the food industry—a review. Trends Food Sci Tech 14:408–423

Weng JZ (2006) Thermal processing of canned foods. In: Sun DW (ed) Thermal food processing, CRC Publications, Boca Raton, p 335

Williamson ME, Wilson DI (2009) Development of an improved heating system for industrial tunnel baking ovens. J Food Eng 91:64–71

Wong SY, Zhou W, Hua J (2006) Robustness analysis of CFD model to the uncertainties in its physical properties for a bread baking process. J Food Eng 77:784–791

Wong SY, Zhou W, Hua J (2007) CFD modeling of an industrial continuous bread-baking process involving U-movement. J Food Eng 78:888–896

Wong SY, Zhou W, Hua J (2007) Designing process controller for a continuous bread baking process based on CFD modeling. J Food Eng 81:523–534

Woo MW, Daud WRW, Mujumdar AS, Talib MZM, Wu ZH, Tasirin SM (2008) CFD evaluation of droplet drying models in a spray dryer fitted with a rotary atomizer. Dry Technol 26:1180–1198

Woo MW, Daud WRW, Mujumdar AS, Talib MZM, Wu ZH, Tasirin SM (2009) Non-swirling steady and transient flow simulations in short-form spray dryers. Chem Prod Process Model 4(1):34, Article 20

Xia B, Sun DW (2002) The application of computational fluid dynamics (CFD) in the food industry: a review. Comput Electron Agric 34:5–24

Xue Z, Ferrua MJ, Singh RP (2012) Computational fluid dynamics modeling of granular flow in human stomach. Alimentos Hoy 21(27):3–14

Zanoni B, Peri C, Bruno D (1995) Modeling of starch gelatinization kinetics of bread crumb during baking. LWT-Food Sci Technol 28:314–318

Zanoni B, Peri C, Bruno D (1995) Modeling of browning kinetics of bread crust during baking. LWT-Food Sci Technol 28:604–609

Zbicinski I (1995) Development and experimental verification of momentum, heat and mass transfer model in spray drying. Chem Eng J 58:123–133

Zbicinski I, Strumillo C, Delag A (2002) Drying kinetics and particle residence time in spray drying. Dry Technol 20(9):1751–1768

Zhang J, Datta AK (2006) Mathematical modeling of bread baking process. J Food Eng 75:78–89

Zhang Z, Zhang W, Zhai Z, Chen QY (2007) Evaluation of various turbulence models in predicting airflow and turbulence in enclosed environments by CFD: Part 2-comparison with experimental data from literature. HVAC&R Res 13(6):871–886

Zhao W, Kumar K, Mujumdar AS (2004) Flow and heat transfer under confined noncircular turbulent impinging jets. Dry Technol 22(9):2027–2049

Zhonghua W, Mujumdar AS (2008) CFD modeling of the gas-particle flow behavior in spouted beds. Powder Technol 183:260–272

Zhou W, Therdthai N (2007) Three-dimensional modeling of a continuous industrial baking process. In: Sun DW (ed) Computational fluid dynamics in food processing. CRC Press, Boca Raton, pp 287–312

Zulkifli R, Sopian K, Abdullah S, Takriff MS (2009) Comparison of local Nusselt number for steady and pulsating circular jet at Reynolds number of 16000. Eur J Sci Res 29:369–378

# Index

C. Anandharamakrishnan, *Computational Fluid Dynamics Applications
in Food Processing*, SpringerBriefs in Food, Health, and Nutrition,
DOI: 10.1007/978-1-4614-7990-1, © Chinnaswamy Anandharamakrishnan 2013